Galina Schrange-Kashenock

Collective Effects in Negative Ion Photodetachment

AF154054

Galina Schrange-Kashenock

Collective Effects in Negative Ion Photodetachment

Many-Body Theory

LAP LAMBERT Academic Publishing

Impressum / Imprint

Bibliografische Information der Deutschen Nationalbibliothek: Die Deutsche Nationalbibliothek verzeichnet diese Publikation in der Deutschen Nationalbibliografie; detaillierte bibliografische Daten sind im Internet über http://dnb.d-nb.de abrufbar.
Alle in diesem Buch genannten Marken und Produktnamen unterliegen warenzeichen-, marken- oder patentrechtlichem Schutz bzw. sind Warenzeichen oder eingetragene Warenzeichen der jeweiligen Inhaber. Die Wiedergabe von Marken, Produktnamen, Gebrauchsnamen, Handelsnamen, Warenbezeichnungen u.s.w. in diesem Werk berechtigt auch ohne besondere Kennzeichnung nicht zu der Annahme, dass solche Namen im Sinne der Warenzeichen- und Markenschutzgesetzgebung als frei zu betrachten wären und daher von jedermann benutzt werden dürften.

Bibliographic information published by the Deutsche Nationalbibliothek: The Deutsche Nationalbibliothek lists this publication in the Deutsche Nationalbibliografie; detailed bibliographic data are available in the Internet at http://dnb.d-nb.de.
Any brand names and product names mentioned in this book are subject to trademark, brand or patent protection and are trademarks or registered trademarks of their respective holders. The use of brand names, product names, common names, trade names, product descriptions etc. even without a particular marking in this work is in no way to be construed to mean that such names may be regarded as unrestricted in respect of trademark and brand protection legislation and could thus be used by anyone.

Coverbild / Cover image: www.ingimage.com

Verlag / Publisher:
LAP LAMBERT Academic Publishing
ist ein Imprint der / is a trademark of
OmniScriptum GmbH & Co. KG
Heinrich-Böcking-Str. 6-8, 66121 Saarbrücken, Deutschland / Germany
Email: info@lap-publishing.com

Herstellung: siehe letzte Seite /
Printed at: see last page
ISBN: 978-3-659-76917-7

Preface

In Brochure the recent development of the Many-Body Theory method, namely RPAE&DEM, will be reviewed. The case of study on photodetachment for a series of negative ions with emphasis on resonant features will be as well presented.

The research group led by Prof. V.K. Ivanov at the St.Petersburg Polytechnical University (Russia) has accumulated experience of many-body calculations for photoprocesses in varied atomic systems. No wonder that it is tackling the problems pertaining to negative ions has given the impetus to the progress of our approach for the consideration of the electron correlation effects. The theoretical study of negative ions could be conducted by the methods elaborated for neutral atoms, but the collective effects in negative ions are often more substantial than those in neutral atoms and sometimes play a dramatic role.

The essence of the new RPAE&DEM approach within the frame of the many-body perturbation theory (Chapter 3 and 6) has been developed in Thesis Work (G.Yu.Kashenock, PhD, St.Petersburg 1998) and presented by the author at the International Workshop on Photoionization (Chester 1997, UK). The latest improvements and applications to a study on the actual problems of the inner-shell negative ions photodetachment (Chapter 10) were reviewed at the Meeting "Fundamental Quantum Processes in Atomic and Molecular Systems" (Sandbjerg 2005, Denmark). The later Chapter 11 presents the newest results on silicon negative ion inner-shell photodetachment and, in some sense, closes the topic by consideration of the most complicated ion in the series of the light negative ions with an open p-shell.

The results of the concrete calculations for the very sensitive to proper account of the many-electrons correlations open-shell negative ions He⁻, Cr⁻, B⁻, C⁻ and Si⁻ and comparison with the available experiments clear demonstrate the high potential of the presented many-body theory method. The objects chosen present the wide range of the resonance types: strong near-threshold and window resonances, giant resonances and interference and autodetachment resonances, shape and Feschbach resonances, and give the most telling example of importance of the collective effects in photodetachment. With using the Feynman-Goldstone diagram technique the physical meaning of the considered collective effects become especially evident.

TABLE OF CONTENTS

1. Introduction

Negative atomic ions are often considered to be perfect objects for fundamental studies of photon-matter interactions within many-particle processes. Investigations of the dynamics of photodetachment processes can provide useful insight to the general problem of correlated motions of electrons in atomic systems, since all negative-ion problems are many-body problems. The collective electron-effects may play a dramatic role since the electron-interaction is taking place in a neutral core field, which is markedly different from the conditions present in neutral atom systems for which the many-electron interactions take place in the unshielded, attractive Coulomb potential. The collective effects are of special interest, when the photodetachment processes occur near the threshold energy or near the resonance energies.

The essence of the new many-body theory method (RPAE&DEM) for simultaneous inclusion of the dynamic polarization potential generated in the system "core + electron", dynamic relaxation (screening) and corrections within the framework of RPAE will be given in the core part of the Brochure. The approach developed uses the Hartree-Fock approximation (HF) as a zero one; many-electron correlations are incorporated within the Random Phase Approximation with Exchange (RPAE, an approximation describing the dynamic collective response of an atomic system on an external field to which the intra- and intershell interactions contribute) (Amusia 1990), the Dyson Equation Method (DEM) (Chernysheva *et al* 1988, Gribakin *et al* 1990, Ivanov *et al* 1996) (and their Spin-Polarized versions). Besides the Many-Body Perturbation Theory (MBPT) methods are applied depending on the role played by individual corrections. With the Feynman-Goldstone diagram technique one may identify certain classes of diagrams from the perturbation expansion and clarify the physical meaning of the included effects. Introducing physically meaningful corrections step by step we are able to analyze their influence on a photodetachment process. To anticipate, it is found that within the scope of RPAE we can make some qualitative conclusions concerning photodetachment channels interference features, but for an adequate description of interference and resonance profiles it is necessary to take proper account of further collective phenomena – dynamical relaxation and polarization.

The finding of an investigation into energy structure, concrete results of calculations of the total and partial cross sections, angular distribution parameters, photoelectron phaseshifts for the He⁻, Cr⁻, B⁻, C⁻, Si⁻ negative ions will be presented. The subjects for study chosen are the ions with open shells. Due to electronic cloud diffusion and possibilities for the formation of quasi-bound states in phototransitions from inner shells to vacant states the important role of collective effects is inherent in photodetachment from these. So there is good reason to believe that the theme provides evidence enough to judge the potential of theoretical approaches for description of electronic correlations.

2. Polarizability, polarization, relaxation

According to the simplest definition "the *photoionization* is the physical process in which an ion is formed from the interaction of a photon with an atom or molecule". When photon interacts with a negative ion and the residual is a neutral atom and photoelectron, we speak about *photodetachment*. The one-electron scheme of the process is given by the upper picture of fig.1: within the single particle approximation we have the electronic structure of our ion (up to Fermi level **F**). From one of the electronic shell (nl') an electron is removed to continuum (ε) when it has received enough energy with a photon ω. The equation for the process of single-photon photodetachment from negative ion **A⁻** could be given as

$$\mathbf{A}^- + \omega \rightarrow \mathbf{A} + \varepsilon e^-.$$

With the spotlight fixed on a variety of collective effects we will give the simplified classification, we will consider three classes of collective effects: *polarizability*, *polarization* and *relaxation*. Below some insight into physical meaning of these effect

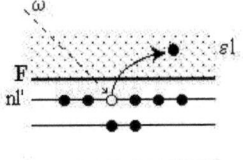

single-particle approximation

POLARIZABILITY

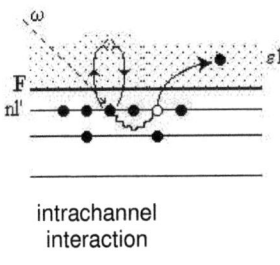

intrachannel
interaction

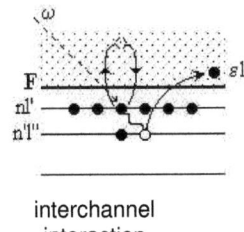

interchannel
interaction

POLARIZATION

RELAXATION

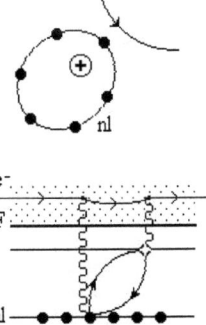

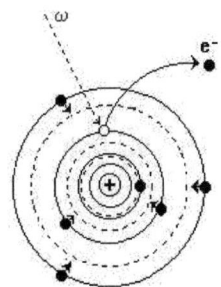

Figure 1

will be given with emphasis to the peculiarities of negative ions. Much of what follows is to be clearer by fig.1.

Photoprocesses dynamic characteristics - total and partial cross sections, photoelectron angular distribution asymmetry parameters - often reflect the fact that the response of an atomic system on an external field is collective (see, for example, Amusia 1990). This is especially true for the atoms (ions) which have a strong channel of photoionization (photodetachment) to interact with other channels. Essential in this case are both electron correlations within the many-electron shell from which the strong transition into continuum comes to pass and intershell interaction. The correlational interaction of this kind, the dynamic polarizability of an atomic system due to external electromagnetic field, and the pertinent consequence of the perturbation theory series (the "lace-like" diagrams with virtual particle-hole pairs) are treated by the RPAE. Intershell interaction is more pronounced with open-shell atomic systems due to existence of quasi-bound states in the vicinity of inner-shell thresholds which may produce resonance structures in the spectrum. For neutral atoms and positive ions the transitions of an inner-shell electron to vacant states of an outer open shell usually are discrete dipole excitations having large oscillator strengths. These transitions shifting into the continuum and manifesting in autoionizing resonances in photoionization cross sections more rarely occur. In negative ions such quasi-bound states typically constitute shape resonances which decay into their own continuum and reveal themselves as strong peaks in the inner shell partial photodetachment cross sections. The intershell correlations, the interaction between the resonance and the phototransitions from the outer shell, cause the photodetachment pattern to depart radically from what is expected from the independent-electron model. An added reason for the considerable impotance of the intershell correlations for negative ions, especially these with open shells, is that the looseness of an electronic cloud (and hence relatively small values of differences in the single-particle energies of an outer and neighboring shells) is characteristic of these systems.

Negative ions structure and photodetachment processes are largely determined by the polarization of a neutral core induced by an extra electron. For adequate description of these, restricting our consideration to the static polarization potential (of the form $V(r) = \frac{\alpha}{2(r^2 + a^2)^2}$, α is a polarizability of neutral atom, a is a parameter) proves to be distinctly deficient. It is necessary that the dynamic, i.e. energy-dependent, core-polarization effects would be taken into account. The dynamic nonlocal polarization potential, which is the irreducible self-energy part Σ of the single-electron Green function of the atom, may be built up *ab initio* within the Simplified RPAE (SRPAE, Chernysheva *et al* 1988). Starting from the HF basis Σ is calculated in the second order of the perturbation theory:

$$\Sigma_E(\vec{p}', \vec{p}) = \quad \text{(1)}$$

Hereafter, the Feynman-Goldstone diagram technique will be used: the line with an arrow to the left (right) represents the hole (particle) which is below (above) the Fermi level **F**; the wavy line is the Coulomb interaction. An additional approximation can be accomplished by using a new set of excited-state wavefunctions for the particle line in the loop: the wavefunction for the m excited states, can be found in the field of an atomic core with a hole j, thus including the interaction of the m electron with the j vacancy. The approximation (1) was formerly applied to the electron-atom low-energy scattering. Without going into details it worth to point out that a succession of expedients is available that permits the inclusion of certain classes of higher order diagrams representing terms of the perturbation series which contribute significantly to self-energy part Σ. One can see that the first diagram of (1) directly corresponds to a schematic description of the polarization process in fig.1. The second diagram is an exchange one; the last couple diagrams (1) are so-called "time-reverse" diagrams in the terms of the Feynman-Goldstone formalism (see for details *e.g.* ref. Amusia 1990).

Another important factor affecting the motion of a photoelectron, especially one detaching from the inner subshell, is a core relaxation. The occurrence of an electron vacancy causes the electronic cloud to deform, rearrange, with the outermost orbit "caving in". The relaxation effects exert the greatest influence at a low-energy photoelectron. If being energetic a photoelectron can rapidly move well away from the nucleus, so that it has no time to perceive the core field change. Then its wavefunction may be calculated in the "frozen" field of an ion with a hole to a good approximation. An another limit is the static relaxation approximation (Generalized RPAE, GRPAE, Amusia 1990). The static relaxation of residual atom is taken into consideration through the calculation of the new HF set of the photoelectron wavefunctions. They are found in the field of the completely rearranged neutral core without an electron in the shell from which the transition into continuum occurs. This approach includes a simplification which is that the core rearrangement due to a hole creation is taken into account right from the instant at which a photon is taken up and photoelectron is ejected, from the start of photoelectron's motion even though it may stay in the interior of an atom for a while. To put this in diagrammatic way, the diagrams corresponding to physically meaningless processes, namely the interaction with a hole before it is created, happen to be included. Until the recent time we have used only this approach. However, when considered within the static approximation the role of relaxation effects often turns out to be overestimated, the photodetachment picture being distorted. For negative ions the fact that the relaxation is a dynamic process is particularly important. An increase of the negative charge density after knocking out a photoelectron proceeds within a neutral system, not in a positive ion as with the photoionization. Hence, as long as a photoelectron is at a short distance from the nucleus, residual electrons strongly "feel" its presence as well as strengthening the nuclear attraction. It follows that the dynamics of core rearrangement process and its influence on the photodetachment must be allowed for – one of the challenge that we will solve in the frame of the RPAE&DEM approach presented below (Chapter 6, 10.1).

3. DEM: calculations and improvement of ground states and polarization influence on final photoelectron state

Consider next the DEM as the method which enables us to bind the outer electron in a negative ion and to correct the wavefunction and the single-particle energy of any electron, inner as well as outer, to take account of the polarization effects on a detaching electron, and thus to consider the influence of polarization interaction between a photoelectron and a core on the photodetachment process.

If the system "neutral atom + electron" is to be considered, the polarisation interaction and exchange interaction between an atom and the extra electron should be taken into account. Starting from the HF basis, the wavefunction $\psi_E(\vec{r})$ describing the motion of the electron with energy E in the atomic field satisfies the Dyson equation:

$$\hat{H}^{(0)}\psi_E(\vec{r}) + \int \Sigma_E(\vec{r},\vec{r}')\psi_E(\vec{r}')d\vec{r}' = E\psi_E(\vec{r}) \tag{2}$$

Here $\hat{H}^{(0)}$ is the static Hartree-Fock Hamiltonian of the atom and $\Sigma_E(\vec{r},\vec{r}')$ is the energy-dependent non-local potential. This equation completely includes the correlational interaction of the extra electron with the atom. $\Sigma_E(\vec{r},\vec{r}')$ is equal to the self-energy part of the single-electron Green function of the atom, hence it can be presented as a diagrammatic expansion in powers of the interelectron correlational interaction.

To define the ground state of a negative ion we follow the method suggested in Chernysheva *et al* (1988), Gribakin *et al* (1990), Ivanov *et al* (1996) and use the Hartree-Fock Hamiltonian $\hat{H}^{(0)}$ spectrum representation. The complete orthonormal set of eigenfunctions $\varphi_\nu(\vec{r})$ which satisfy

$$\hat{H}^{(0)}\varphi_\nu(\vec{r}) = \varepsilon_\nu\varphi_\nu(\vec{r}) \tag{3}$$

includes wavefunctions of discrete states ($\nu = nl$) and continuum states ($\nu = \varepsilon l$), which describe the electron of energy ε and orbital momentum l, scattered by the Hartree-Fock potential of the neutral atom. Using wavefunctions defined by equation (3) as a basis the binding energy and the wavefunction of an electron may be found by solving equation (2) like the eigenvalue problem for a discrete spectrum (Chernysheva *et al* 1988). My contribution to development of the DEM is modification of the method

11

which makes it possible to define the new corrected energy and wavefunction for an electron state bound already within the HF potential, and working out its Spin-Polarized version (Ivanov *et al* 1996a, Ivanov *et al* 1996b).

Solving equation (2) one may obtain the wavefunction of an electron state with new eigenvalue E_i, which can be generally represented diagrammatically as:

$$(4)$$

Hereafter the line marked with an arrow in the form of close triangle corresponds to the DEM-wavefunction and line free of an arrow can represents both the hole ($\varepsilon_v \leq F$) and the particle ($\varepsilon_v > F$) states. The diagrams of the first order of perturbation theory (4b, 4c), diagrams of the HF type, are included in the case that E_i l is an inner subshell and the outer electron (E_A) in the negative ion cannot be bound within the HF approximation. In this situation it is necessary to take into account diagrams which describe Coulomb interaction of electrons in the HF configuration (netral atom) with outer electron of the ion. Within the Spin-Polarized version the exchange diagrams 4c (or 4e,4g) should be omitted provided that E_i l and E_A (or E_i l and j, respectively) are electron states of opposite spin projections.

To account for the influence of the atom-core dynamic polarization potential on an outgoing electron the following effective dipole amplitudes, which include the irreducible self energy part Σ of the single-particle Green's function, should be considered:

$$l = l' \pm 1 \qquad (5)$$

Here $\varepsilon_0 = \omega + \varepsilon_i$ is an energy parameter, ω is a photon energy (atomic units are used), ε_i is the initial state energy of the photodetachment channel in which allowance is made for the polarization effects. $\tilde{\Sigma}$ is the reducible self energy part of single-particle Green's function which is a solution of the following integral equation in diagrammatic representation:

$$(6)$$

Analytically the effective dipole amplitude (5) and equation (6) may be written as:

$$\langle \varepsilon | \hat{D}_{\varepsilon_0} | \tilde{\imath} \rangle = \langle \varepsilon | \hat{d} | \tilde{\imath} \rangle + \int_{\varepsilon'} \frac{\langle \varepsilon | \hat{\Sigma}_{\varepsilon_0} | \varepsilon' \rangle \langle \varepsilon' | \hat{d} | \tilde{\imath} \rangle}{\varepsilon_0 - \varepsilon' + i\delta} + \int_{\varepsilon'\varepsilon''} \frac{\langle \varepsilon | \hat{\Sigma}_{\varepsilon_0} | \varepsilon'' \rangle \langle \varepsilon'' | \hat{\Sigma}_{\varepsilon_0} | \varepsilon' \rangle \langle \varepsilon' | \hat{d} | \tilde{\imath} \rangle}{(\varepsilon_0 - \varepsilon'' + i\delta)(\varepsilon_0 - \varepsilon' + i\delta)} + ... =$$

$$= \langle \varepsilon | \hat{d} | \tilde{\imath} \rangle + \int_{\varepsilon'} \langle \varepsilon' | \hat{d} | \tilde{\imath} \rangle \left(vp \frac{1}{\varepsilon_0 - \varepsilon'} - i\pi\delta(\varepsilon_0 - \varepsilon') \right) \Big\{ \langle \varepsilon | \hat{\Sigma}_{\varepsilon_0} | \varepsilon' \rangle +$$

$$+ \int_{\varepsilon''} \langle \varepsilon'' | \hat{\Sigma}_{\varepsilon_0} | \varepsilon' \rangle \left(vp \frac{1}{\varepsilon_0 - \varepsilon'} - i\pi\delta(\varepsilon_0 - \varepsilon') \right) \{ \langle \varepsilon | \hat{\Sigma}_{\varepsilon_0} | \varepsilon'' \rangle + ... \} \Big\} = \qquad (5a)$$

$$= \langle \varepsilon | \hat{d} | \tilde{\imath} \rangle + vp \int_{\varepsilon'} \frac{\langle \varepsilon | \hat{\tilde{\Sigma}}_{\varepsilon_0} | \varepsilon' \rangle \langle \varepsilon' | \hat{d} | \tilde{\imath} \rangle}{\varepsilon_0 - \varepsilon'} - \frac{i\pi \langle \varepsilon | \hat{\tilde{\Sigma}}_{\varepsilon_0} | \varepsilon_0 \rangle}{1 + i\pi \langle \varepsilon_0 | \hat{\tilde{\Sigma}}_{\varepsilon_0} | \varepsilon_0 \rangle} \left(\langle \varepsilon_0 | \hat{d} | \tilde{\imath} \rangle + vp \int_{\varepsilon'} \frac{\langle \varepsilon_0 | \hat{\tilde{\Sigma}}_{\varepsilon_0} | \varepsilon' \rangle \langle \varepsilon' | \hat{d} | \tilde{\imath} \rangle}{\varepsilon_0 - \varepsilon'} \right)$$

13

$$\langle \varepsilon''|\hat{\tilde{\Sigma}}_{\varepsilon_0}|\varepsilon'\rangle = \langle \varepsilon''|\hat{\Sigma}_{\varepsilon_0}|\varepsilon'\rangle + vp\int_{\varepsilon_v} \frac{\langle \varepsilon''|\hat{\tilde{\Sigma}}_{\varepsilon_0}|\varepsilon_v\rangle\langle \varepsilon_v|\hat{\Sigma}_{\varepsilon_0}|\varepsilon'\rangle}{\varepsilon_0 - \varepsilon_v} \tag{6a}$$

Where $\hat{d}$ is the single-particle dipole operator in the length (r-) or velocity (∇-) form, $\delta(\varepsilon)$ is the Dirac delta-function and vp is the principal value symbol. Matrix elements are determined through the use of the HF wave functions $|v\rangle$ and the ground state wavefunction corrected within the DEM $|\tilde{i}\rangle$ by equation (4); we often drop some quantum numbers of the set $v = (n(\varepsilon), l, m, \mu)$ for the sake of brevity. Throughout this paper $\int_{\varepsilon_v}$ denotes the summation and integration over the whole spectrum of v states, occupied as well as excited.

It should be noted that for the amplitudes D_{ε_0} on the energy shell ($\varepsilon_0 = \varepsilon$) expression (5a) may be reduced to a simpler one. If the first two summands in the last line of (5a) are identified with the first term a_0 of a geometric progression and if $q = \langle \varepsilon|\hat{\tilde{\Sigma}}_\varepsilon|\varepsilon\rangle$ then the expression for D_ε may be written in the form:

$$a_0 + q(a_0 + qa_0 + q^2a_0 + ...) = a_0 + qa_0 + q^2a_0 + ... = \frac{a_0}{1-q} \tag{7}$$

That is to say, with the constraint $\varepsilon_0 = \varepsilon$, i.e. the state of energy ε is a real final state for the photodetachment process, the infinite sequence of diagrams (5) gives a geometric series which is readily (Ivanov *et al* 1996) summed to

$$\langle \varepsilon|\hat{D}_\varepsilon|\tilde{i}\rangle = \frac{1}{1 + i\pi\langle \varepsilon|\hat{\tilde{\Sigma}}_\varepsilon|\varepsilon\rangle}\left(\langle \varepsilon|\hat{d}|\tilde{i}\rangle + vp\int_{\varepsilon'} \frac{\langle \varepsilon|\hat{\tilde{\Sigma}}_\varepsilon|\varepsilon'\rangle\langle \varepsilon'|\hat{d}|\tilde{i}\rangle}{\varepsilon - \varepsilon'}\right) \tag{8}$$

Equation (8), an analogue of the Dyson equation, can be applied to taking account of dynamic polarization effects on a photodetachment channel which is scarcely affected by other channels, ignoring intershell interaction.

4. 2p,2s-photodetachment from He⁻

The approach as stated above has been taken to a simple but interesting system - the He⁻ negative ion. The 1s2s2p ($^4P^o$) metastable state of He⁻ is a classic example of a

Feshbach resonance (the lifetime of the order of 10^{-5} s). Although the ion's electrons are few in number, account must be taken of the many-electron correlations if the photodetachment processes, specifically effects associated with influence of the well-known $1s2p^2$ ($^4P^e$) shape resonance, are to be described adequately.

Opening up a new subshell the outer p-electron cannot be bound within the HF approximation because this single-particle method neglects the attractive polarization interaction. Chernysheva *et al* (1988) have applied the DEM to calculation of ground state wavefunctions and to the 2p He⁻ photodetachment, where the electron affinity for the He(2^3S) has been found to be equal *0.084 eV* and the photodetachment cross sections for 2p→ εs,εd channels were calculated for a photon energy up to *4 eV* with Hartree-Fock (HF) wavefunctions for the final states. Recently we have reported (Ivanov *et al* 1996a) the results of improved calculations for the wavefunctions in initial 2p (ε_{2p}= -80 meV, $E_A^{exp}(2^3S)$= 77.67±0.12 meV (Walter *et al* 1994)) and final εd,εs states (by taking into account additional excitations in the solution of the Dyson equation and the action of atom-core polarization on the photoelectron) and also the first calculations for 2s photodetachment within the DEM.

The photodetachment cross sections for the 2p→es and 2p→ed channels calculated with the self-energy part taken into account and total cross section (in r-form) are shown in the fig.2. The position and value of the total cross section maximum are in gratifying agreement with the experimental results (Hodges *et al* 1981), especially as the authors found an absolute uncertainty of 30% had to be assigned to the data. The results on the photoelectron angular distribution asymmetry parameter β_{2p}, obtained using the usual formula (Amusia 1990), together with the experimental data (Thompson *et al* 1990) are presented in fig.3.

The most interesting feature in He⁻ photodetachment, with which is our prime interest here, is the strong narrow resonance just after the 2s threshold, associated with the shape resonance in the 2s→ εp (^4P) transition to the continuum spectrum. The latter is due to the "$1s2p^2$" (^4P) state of He⁻, which appears to lie just above its parent He (2^3P) (Bunge and Bunge 1979, Nicolaides *et al* 1981). The resonance effects proved to be

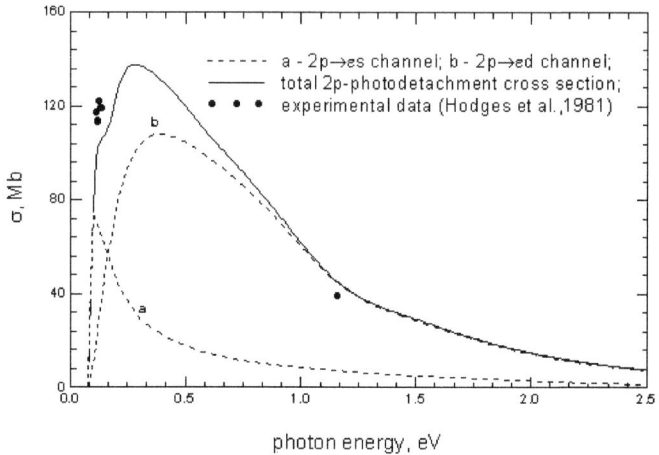

Figure 2. The 2p-photodetachment cross section for He$^-$ $1s2s2\tilde{p}$ ($^4P^o$) (in a length form of the dipole operator). The calculations were performed with the polarization potential (Σ) taken into accont. The experimental data are depicted only for a photon energy up to the 2^3P threshold.

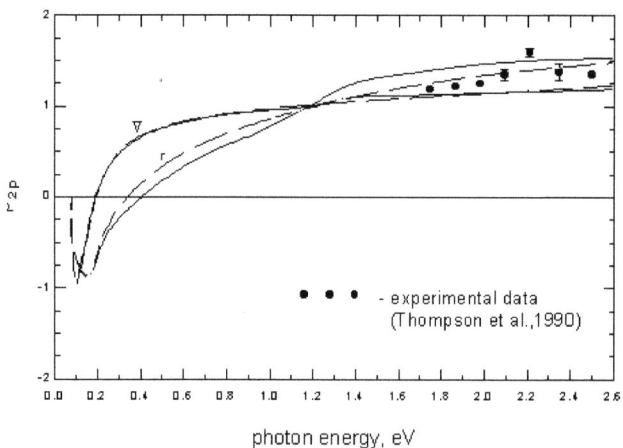

Figure 3. The photoelectron angular asymmetry parameter for 2p He$^-$ ($^4P^o$). Dashed and solid curves are results within the HF approximation and with the polarization potential (Σ) taken into accont, respectively.

obtainable by consideration of virtual states of the "atom-core + photoelectron" system in calculating the irreducible self-energy part Σ. Correlational interaction of a photoelectron in the εp continuum with electrons of the He (2^3P) core results in the "temporary negative ion" $1s2p^2$ ($^4P^e$) formation. Polarization of the atom by the outgoing electron causes a dynamic attractive potential which is supplementary to the static short-range attractive potential generated by the atom. The attraction is not strong enough to form a Feshbach resonance but the angular momentum of the electron ($l=1\neq0$) forms a penetrable centrifugal barrier and the electron temporarity unites with the atom to yield a shape resonance.

Any treatment of 2s-photodetachment for the He⁻ would be impossible without an improvement on the description of 2s-electron behaviour in the ion, i.e. more exact definitions of the 2s wavefunction and single-particle energy, because the experimental 2^3P threshold of the He⁻ ion $\varepsilon_{2s}^{exp} = -1.222\ eV$ (Walter $et\ al$ 1994) differs significantly from the neutral He HF value $\varepsilon_{2s}^{HF} = -4.74\ eV$ which was ascribed in the starting approximation. Using the Dyson equation we have obtained the corrected owing to consideration of the Coulomb and polarization interaction with the outer 2p-electron 2s wavefunction and its energy $\varepsilon_{2s} = -1.07 eV$. Once the 1s wavefunction has also be corrected we have for the total energy difference $E^{tot}(He^-) - E^{tot}(He\ 2^3P^o) = -1.17 eV$ and one can see that this value is quite close to the experimental 2^3P threshold.

The improved 2s wave function was used for the photodetachment cross section calculations both with and without the account of polarization effects on outgoing electron. Without the polarization influence the 2s shape resonance is too far from the threshold and has a small peak value. The account of the polarization potential influence leads to the narrow shape resonance with the correct position at the energy $\omega = \omega_{res} = 1.233 eV$ (Walter $et\ al$ 1994) and with peak value about 3.5×10^{-15} cm^2 (see fig.4). This value agrees more closely with the range $(5.8\pm2.0)\times10^{-15}$ cm^2 resulting from Walter $et\ al$'s experiment than do the theoretical values obtained by other means $(2.4\times10^{-15}$ cm^2, extensive configuration-interaction wavefunctions and the Stieltjes-moment-theory technique (Hazi and Reed 1981) and ~10^{-16} cm^2, the Multiconfiguration

Hartree-Fock method (Saha and Compton 1990)). Integrating our theoretical cross section over the resonance region from threshold to *1.25 eV* gives the resonance strength as an optical transition *0.42* (r-form) and *0.56* (∇-form) which values are very close to $f_{2s \to 2p} = 0.539$, the oscillator strength of the 1s2s (^{3}S)$\to$1s2p (^{3}P) bound-bound transition in neutral He.

The shape resonance 1s2p^2 (^{4}P) acts strongly on the threshold 2s cross section, that agrees well with experimental data and the Peterson *et al*'s (1985) parametric formula for the behaviour of opening-channel cross sections in the vicinity of a shape resonance (it is a product of the Wigner threshold law for an opening p-wave channel and the Breit-Wigner resonance formula:

$$\sigma \propto \frac{\left(\omega - E_{thr}\right)^{3/2}}{\left(\omega - \omega_{res}\right)^2 + \left(\dfrac{\Gamma}{2}\right)^2}, \tag{9}$$

here ω is the photon energy, E_{thr} is the threshold energy of the channel, ω_{res} is the energy of the resonance, Γ is the decay width of the resonance).

The radial part of the photoelectron wavefunction satisfying the Dyson equation has the asymptotic form (energy is measured in rydbergs):

$$P_{\varepsilon l}^{\Sigma}(r) \propto \frac{1}{\sqrt{\pi\sqrt{\varepsilon}}} (\sin(\sqrt{\varepsilon}r - \frac{\pi l}{2} + \delta_l^{HF}(\varepsilon)) -$$
$$-tg\Delta\delta_l(\varepsilon)\cos(\sqrt{\varepsilon}r - \frac{\pi l}{2} + \delta_l^{HF}(\varepsilon))), \quad r \to \infty \tag{10}$$

where $\Delta\delta_l(\varepsilon)$2 is the correlational correction to the HF value $\delta_l^{HF}(\varepsilon)$3 of the phaseshift ($P_{\varepsilon l}^{HF}(r) \propto \frac{1}{\sqrt{\pi\sqrt{\varepsilon}}}\sin(\sqrt{\varepsilon}r - \frac{\pi l}{2} + \delta_l^{HF})$, $r \to \infty$4). From equations (2), (6), (10) one can find

$$tg\Delta\delta_l(\varepsilon) = -\pi\langle\varepsilon|\tilde{\Sigma}_{\varepsilon}|\varepsilon\rangle, \tag{11}$$

the expression for this correction (Amusia *et al* 1975).

In fig.5 is the p-wave phaseshift for the 2s$\to$$\varepsilon$p (^{4}P) photodetachment channel within the HF approximation and with correlational corrections taken into account is shown. The phaseshift behaviour indicates a formation of added attractive potential ($\Delta\delta_l > 0$5). The total phaseshift is divisible into two parts a slowly changing background and a

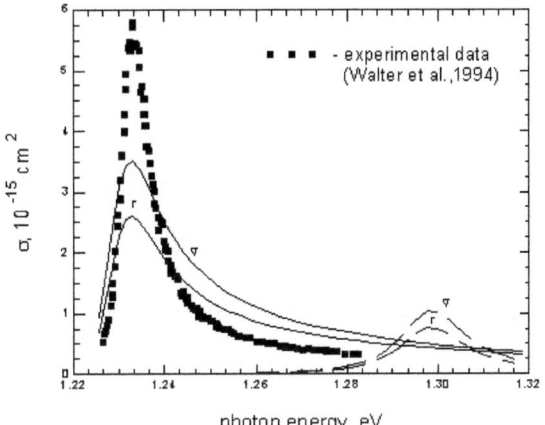

Figure 4. The 2s-photodetachment cross section for He⁻ $1s2\tilde{s}\,2\tilde{p}$ ($^4P^o$). The channel:

$$He^-\,(^4P)+\omega \rightarrow He(2^3P)+e^-\,(\varepsilon p)\ ^4P.$$

The calculation are performed using the corrected $2\tilde{s}$-wavefunction and the experimental 2s binding energy value $E_{2s}=1.222\,eV$ and include the polarization potential influence upon the outgoing electron. The dashed curve is the result within the HF approximation.

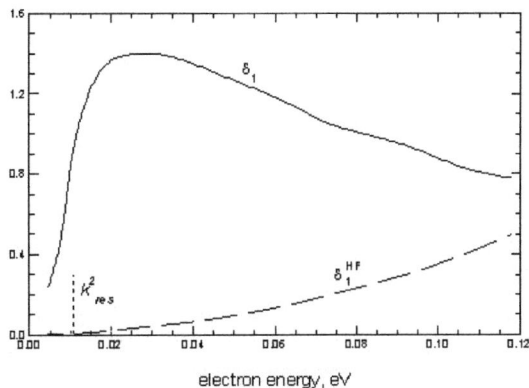

Figure 5. The scattering ^4P phaseshift for an εp-photoelectron. The phaseshift $\delta_1(\varepsilon)$ (solid curve) is presented as a sum of the HF phaseshift $\delta_1^{HF}(\varepsilon)$ (dashed curve) and an additional phaseshift $\Delta\delta_1(\varepsilon)$ due to the polarization potential (Σ): $\delta_1(\varepsilon)=\delta_1^{HF}(\varepsilon)+\Delta\delta_1(\varepsilon)$. The phaseshift $\delta_1(\varepsilon)$ demonstrates the existence of the $1s2p^2$ ($^4P^e$) shape resonance at the electron energy k_{res}^2.

resonance one which rapidly rises over a narrow energy region. Fig.5 provides a further method of resonance parameter determination.

Hence the role of a polarization effect on photoelectron was shown to be significant. It was found that polarization effects are responsible for the formation of the $1s2p^2$ ($^3P^o$) shape resonance. But in view of the weakness of the $2s \to \varepsilon p$ (^4S and ^4D) photodetachment channels, which and only which can interact with 2p-photodetachment channels, the intershell interaction was negligible in these calculations.

The results in details are presented in [1] (references in Appendix II).

5. 4s-photodetachment from Cr⁻

Many-body calculations for outer 4s-subshell photodetachment of the negative ion Cr⁻ were performed as well. The 4s-cross section reveals a near threshold strong maximum which may be attributed to the presence of a quasi-bound "4p" state in the continuum.

Because of the fact that the neutral Cr atom has half-filled 3d and 4s subshells, it is most convenient to use the Spin-Polarized version of the HF approximation which treats half-filled subshells as closed. Then within the SPHF the electronic structure of Cr⁻ can be presented as:

$$1s\uparrow 1s\downarrow....3p^3\uparrow 3p^3\downarrow 3d^5\uparrow 4s\uparrow 4s\downarrow \quad (^6S),$$

where the arrows denote the electron spin projection. However, the $4s\downarrow$ electron cannot be bound within the SPHF method. In Ivanov *et al* 1996b *ab initio* calculations have been performed in which the $4s\downarrow$ wave function was obtained by the solving of the Dyson equation. The energy-dependent nonlocal polarization potential of the Cr atom for $4s\downarrow$ electron Σ was calculated in the second order of perturbation theory including the monopole, dipole, quadrupole and octupole excitations of the $4s\uparrow$ and $3d\uparrow$ electrons. The electron affinity of Cr obtained (0. 782 eV) is close to experimental one (0.667 eV, Feigerle *et al* 1981). The inclusion of the Coulomb interaction with the extra $4s\downarrow$ electron and second-order corrections caused the energy structure of the ion to be substantially refined relative to the structure of the ground state obtained by adding of 4s↓ electron to the SPHF neutral Cr core (see table 1).

Table 1. The single-electron binding energies for Cr⁻ (in eV). The experimental data are from Radtzig and Smirnov (1986), the SPHF energies were obtained for the neutral Cr atom, the DEM values are calculated using the equation (4).

nlμ	Experiment	SPHF	DEM
3p↑	-	61,5	45.6
3p↓	-	50.0	36.0
3d↑	1.66	10.69	1.49
4s↑	1.50	6.04	1.25
4s↓	0.667	-	0.782

Knowledge of the reducible self-energy part $\tilde{\Sigma}$ permits the determination of the l-electron scattering phaseshifts $\delta_l(\varepsilon) = \delta_l^{HF}(\varepsilon) + \Delta\delta_l(\varepsilon)$ from equation (11). The phaseshifts for the s- and p-wave scattering obtained within both the SPHF and the DEM as a function of electron energy are presented in fig.6. One can see that the dynamic polarization potential changes substantially the behaviour of the scattered s-wave phase. The phase shift arising is equal π at zero electron energy. According to Levinson's theorem it means that a new discrete s-level has arisen in the system (Cr + e⁻), i.e. the negative ion Cr⁻ with electron configuration ...3d⁵↑4s↑4s↓ exists. The dynamic polarization potential influences on the p-wave scattering less than the s-wave and its presence does not lead to the binding of the p-wave electron to the neutral Cr atom. However the polarization potential significantly effects both the continuum wavefunctions and the photodetachment cross section.

The results of the 4s↓ photodetachment cross section calculations are presented in fig.7. One can see the typical maximum just after the 4s↓ threshold, which is associated with the quasi-bound resonance "4p" state in the εp↓ continuum. The DEM has been used to account for the polarization effects on both the 4s↓ electron and the εp↓ photoelectron. The influence of the polarization potential shifts the

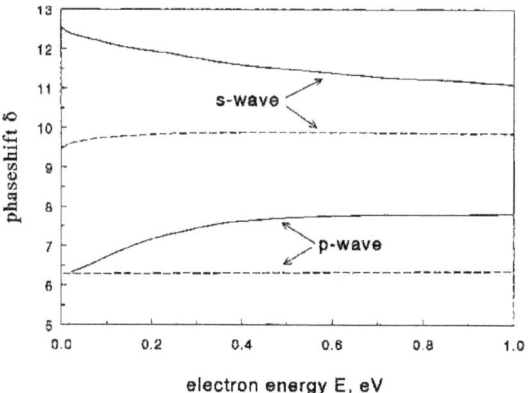

Figure 6. s-wave and p-wave electron scattering phaseshifts, obtained within the SPHF approximation (dashed line) and with including of the polarization interaction (solid line)

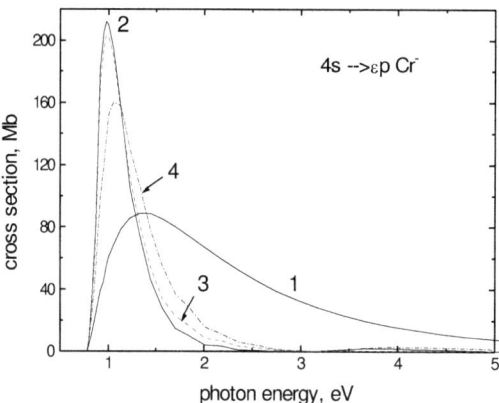

Figure 7. The 4s↓ photodetachment cross section with polarization effects taken into account. 1 - with the SPHF εp wave functions (length form); 2,3 - with the εp wave functions, obtained within the DEM (with account for 17 transitions from from 4s↑ and 3d↑ subshells in the self-energy calculations) in length and velocity forms, respectively; 4 - with the εp wave functions, obtained within the DEM with 5 transitions from 4s↑ subshell only included.

"4p↓" resonance in the $\varepsilon p\downarrow$ continuum closer to threshold and significantly increases the peak value (from 88 Mb up to 213 Mb in length form). Comparison between the curves 2 and 4 in fig.7 shows that even a relatively weak contribution of excitations of the 3d↑ deep-seated subshell to the self-energy part results in a change in the cross section. Hence it is shown that the resonance depends strongly on core polarization effects.

The results in details are presented in [2] (references in Appendix II)

6. RPAE&DEM:
many-body theory method for concurrent consideration
of intra- and interchannel interaction
and dynamic core polarization and relaxation

When the many-body theory is applied to investigation of systems with inherent strong intra- and interchannel interaction, the simultaneous inclusion of correlation within the framework of the RPAE and other significant collective effects becomes a principle problem.

The integration of the DEM and RPAE was a problem posed as main issue of the Thesis Work (Kashenock 1998). The bacic challenge can be boiled down to the following schematic skech:

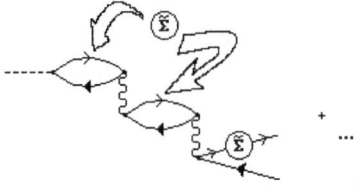

it involves taking into account polarization influence on excited electron in virtual intermediate states. The photodetachment process is expressed graphically as repeated interaction of electrons with the atom-core (represented by Σ (see equation (1)) in the particle lines of diagrams) interspersed with excitations of virtual electron-hole pairs.

Since the relaxation of residual atom proved to be significant (see concrete examples below) the problem was augmented to include this effect. Static relaxation can be considered within our method without adding complexity to the computer program algorithms, but for a dynamic character of the process to be taken into account a new advanced approach was called for.

An approach for simultaneous inclusion of the dynamic polarization, dynamic relaxation of a residual atom (screening) and RPAE correlations has been elaborated. It is realized as a computer code for calculation of dipole transition amplitudes and photodetachment (photoionization) cross sections and photoelectron angular distribution asymmetry parameters.

Taking account of the dynamic polarization effects is achieved by consideration of effective dipole and Coulomb amplitudes. They must be calculated off as well as on the energy-shell. With this in mind the general equation for D_{ε_0} (5a) has been derived. In just the same way consideration must be given to polarization corrections to Coulomb matrix elements. Summation of an infinite sequence of diagrams which is akin to (5) gives an expression for the effective Coulomb amplitudes:

$$\langle \varepsilon \tilde{j} | \hat{U}_{\varepsilon} | E \tilde{i} \rangle = \qquad (12)$$

Which may be written analytically in the following way

$$\left\langle \varepsilon \tilde{j} | \hat{U}_{\varepsilon_0} | E \tilde{i} \right\rangle = \left\langle \varepsilon \tilde{j} | \hat{u} | E \tilde{i} \right\rangle + vp\!\int_{\varepsilon'}\frac{\left\langle \varepsilon | \hat{\tilde{\Sigma}}_{\varepsilon_0} | \varepsilon' \right\rangle\!\left\langle \varepsilon' \tilde{j} | \hat{u} | E \tilde{i} \right\rangle}{\varepsilon_0 - \varepsilon'} - $$
$$-\frac{i\pi\left\langle \varepsilon | \hat{\tilde{\Sigma}}_{\varepsilon_0} | \varepsilon_0 \right\rangle}{1 + i\pi\left\langle \varepsilon_0 | \hat{\tilde{\Sigma}}_{\varepsilon_0} | \varepsilon_0 \right\rangle}\left(\left\langle \varepsilon_0 \tilde{j} | \hat{u} | E \tilde{i} \right\rangle + vp\!\int_{\varepsilon'}\frac{\left\langle \varepsilon_0 | \hat{\tilde{\Sigma}}_{\varepsilon_0} | \varepsilon' \right\rangle\!\left\langle \varepsilon' \tilde{j} | \hat{u} | E \tilde{i} \right\rangle}{\varepsilon_0 - \varepsilon'}\right) \qquad (12a)$$

Here $\langle \varepsilon \tilde{j} | \hat{u} | E \tilde{i} \rangle = \langle \varepsilon \tilde{j} | \hat{v} | E \tilde{i} \rangle - \langle \tilde{j} \varepsilon | \hat{v} | E \tilde{i} \rangle \cdot \delta_{\mu_i \mu_j}$ are the single-particle antisymmetrized Coulomb matrix elements, where $\hat{v} = \dfrac{1}{|\vec{r} - \vec{r}'|}$ (atomic units) and the factor $\delta_{\mu_i \mu_j}$ cancels the exchange interaction of electrons with different spin projections μ. However, starting from the RPAE which considers the many-electron correlations that come from multiple pair interactions of the type "two particle - two hole", we restrict the integration (summation) over ε' in (12a) to excited states, i.e. omit the diagrams bracketed in (12).

The effective amplitudes D_{ε_0} and U_{ε_0} are substituted into the RPAE equations for the single-particle ones. On solving a system of these equations we find the dipole transition amplitudes with allowance made not only for the intra- and interchannel correlations, but for the effect of the dynamic core-polarization upon electrons in intermediate and final states as well. Thus, the refined RPAE equations take the form:

$$(13)$$

Because the ground state wavefunctions are corrected by solving equation (4), the Σ-blocks may be thought of as being inserted into the hole lines too.

As things now stand, we exclude the influence of the polarization potential in the processes described by "time-reverse" diagrams from consideration, because electrons in the virtual time-backward intermediate states move in a field of an ion with two holes. The existence of one more hole in no way affects the calculation of electron wavefunctions, so that there is little or no point in taking into account corrections (1) in the second and higher orders of perturbation theory.

For every photodetachment channel $i_k l'_k \to \varepsilon l_k$ the continuous spectrum of excited states is approximated by a finite set of HF $\varepsilon_\nu^{(k)} l_k$ states which are equidistant in momentum (Amusia and Chernysheva 1983). Sets of wavefunctions stored on a disk are used in the calculation of dipole and Coulomb matrix elements $\langle \varepsilon_\nu^{(k)} | \hat{d} | \tilde{i}_k \rangle$, $\langle \varepsilon_{\nu_1}^{(k_1)} \tilde{i}_{k_2} | \hat{u} | \varepsilon_{\nu_2}^{(k_2)} \tilde{i}_{k_1} \rangle$. These are then used in the system of linear algebraic equations which replaces the integral ones within the method of numerical solution of RPAE equations (Amusia and Chernysheva 1983) that are followed here. If the polarization corrections are to be considered for certain of the channels ($k *$ for one), the corresponding effective amplitudes must substitute for the single-particle ones. To determine them, for each of energy parameter values $\varepsilon_0 = \varepsilon_\nu^{(k)} - \varepsilon_{i_k} + \varepsilon_{i_{k*}}$ we build up the self-energy matrix $\langle \varepsilon_{\nu'}^{(k*)} | \hat{\Sigma}_{\varepsilon_0} | \varepsilon_{\nu''}^{(k*)} \rangle$ with the same set of wavefunctions for the channel of photodetachment into the l_{k*}-continuum and then solve equation (6a) for the reducible self-energy part matrix elements $\langle \varepsilon_{\nu'}^{(k*)} | \hat{\tilde{\Sigma}}_{\varepsilon_0} | \varepsilon_{\nu''}^{(k*)} \rangle$. The matrix elements of $\hat{d}$, $\hat{u}$, $\hat{\tilde{\Sigma}}_{\varepsilon_{0-}}$ in (5a) and (12a) corresponding to transitions from and to the state of energy ε_0 (if it is not equal to any $\varepsilon_\nu^{(k)}$ of the set chosen and the wavefunction is not known) are computed by a linear interpolation procedure. Since the integrands have a pole, numerical solution of the equations involves integration in the principal value sense. We follow the numerical methods described by Amusia and Chernysheva (1983). It is worth noting that the effective amplitudes D_{ε_0} and U_{ε_0} are generally complex, whereas the single-particle ones d, u are always real-valued. The appearance of nonzero imaginary parts in the dipole and Coulomb matrices and the photon energy (input energy) dependence of the matrix elements cause the algorithm to solve the RPAE equations to become somewhat more complicated. Corresponding FORTRAN codes were written for a PC.

To incorporate the dynamic core-relaxation effects one further type of diagrams comprising the mass operator Σ contained in a kernel of the Dyson integral equation (6a) should be included. The following equation

$$(14)$$

is solved and in doing so the screening interaction matrix elements are found. Then the new effective amplitudes on and off the energy-shell are calculated from (5a) and (12a).

The term "screening" needs clarification. Let us consider the effective dipole amplitudes which obey the equation:

$$(15)$$

As already noted, relaxation is important for a photoelectron which escapes from an inner i subshell. Photoelectron wavefunctions being calculated in the "frozen-core" field of an ion with a i hole and with the definite total system symmetry, the first term of equation (15) includes the hole field (Amusia 1990). It incorporates the subsequence of "time-forward" RPAE diagrams in which the hole state remains unchanged in the energy:

$$(16)$$

Here N+1 indicates that the function is found in the field of an ion. The final pair of terms of equation (15) represent an infinite sequence of diagrams which describe virtual monopole excitations of a j (outer) electron. To include the contribution of these diagrams is to allow for the fact that the hole field weakens due to screening. So the interaction between a photoelectron and a hole is dynamically screened by the other electrons of the residual atom with the outer shell contribution being the most pronounced. By choosing a set of wavefunctions for j and k states one can also consider the interaction of particle and hole lines of the loop with the i electron vacancy. The exchange terms are absent in equation (15) since nowadays we deal only with spin-polarized neighboring subshells, which contain electrons with opposite spin projections. The dipole matrix elements with allowance made for dynamic core relaxation (denoted by closed circles), the new effective photodetachment amplitude, comprises a sequence of diagrams with any number of screening loops. The first terms (two last diagrams in (15) without closed circles in the vertices) introduce a correction which is of opposite sign and can be comparable to the HF dipole amplitude in value. Since it alternates, the series must be summed.

Both of the above corrections, dynamic relaxation and polarization, may be considered simultaneously by combining equations (6), where the Σ-blocks are defined as diagrams (1), and (14).

Substituting the obtained from (13) dipole amplitudes and HF electron scattering phases into usual formulae (Amusia 1990, Sobel'man 1963) one can define the photodetachment cross sections and photoelectron angular distribution.

7. B⁻ photodetachment

The approach developed was taken at calculations of the partial and total photodetachment cross sections for the B⁻ negative ion (Kashenock and Ivanov 1997). The B⁻ photoabsorption spectrum is intriguing due to strong interaction between direct transitions of the outer 2p-electron into continuum and transition of the 2s-electron to a quasi-bound state in the open 2p-subshell.

The extra 2p-electron in the B⁻ negative ion can be bound within the SPHF to form the triplet B⁻...2s↑2s↓2p↑² (³P) ground state. However, the calculated 2p single-electron energy (*0.72 eV*) differs significantly from the experimental electron affinity for the B (*0.28 eV*, Feigerle *et al* 1981). The value obtained within the DEM with the monopole, dipole and quadrupole excitations of the 2p↑ and 2s↓ subshells taken into account in calculating the irreducible self-energy part Σ is *0.27 eV*. The improved 2p wavefunction and the SPHF term-dependent excited state wavefunctions were used in photodetachment cross section calculations within the SP RPAE to take into account the intra- and interchannel interactions. Although the RPAE method was developed to deal only with atoms and ions with filled or half-filled subshells (Amusia 1990) and the B⁻ negative ion has an incomplete spin-polarized 2p↑-subshell being more difficult to treat with this method, the open-shell problem has been reduced to the consideration of three-electron configurations. The distinctive characteristic of the photodetachment process is the strong interaction of the 2p↑→εd↑ (³P, ³D) continua and the "2s↑2p²↑2p↓" ³P and ³D shape resonances in the unperturbed (2s↑2p²↑+εp↓) continua, the latter being most pronounced. When taking into account intershell correlations, the interaction of the 2s↓→"2p↓" ³P and ³D resonances with 2p↑→εs↑, εd↑ (³P) and 2p↑→εd↑ (³D) channels leads to the interference. However when the prediction based on the SP RPAE have been compared with the experimental evidence it became apparent that for adequate description of the B⁻ photodetachment process it is necessary to go beyond the RPAE method. Although the interchannel interaction exerts primary control over the photodetachment process, proper allowance must be made for further many-electron correlations.

Within the new method the following channels for photodetachment from the outer subshells have been considered:

I. The $^3P \rightarrow {}^3P^o$ total system symmetry transitions

$$... 2p^2\uparrow \ (^3P) + \omega \rightarrow ... 2p\uparrow \ (^2P^o) + \varepsilon s\uparrow \ (^3P^o)$$
$$... 2p^2\uparrow \ (^3P) + \omega \rightarrow ... 2p\uparrow \ (^2P^o) + \varepsilon d\uparrow \ (^3P^o)$$
$$... 2s\downarrow 2p^2\uparrow \ (^3P) + \omega \rightarrow ... 2p\uparrow^2 \ (^4P) + \varepsilon p\downarrow \ (^3P^o)$$

II. The $^3P \rightarrow {}^3D^o$ total system symmetry transitions

$$... 2p^2\uparrow \ (^3P) + \omega \rightarrow ... 2p\uparrow \ (^2P^o) + \varepsilon d\uparrow \ (^3D^o)$$
$$... 2s\downarrow 2p^2\uparrow \ (^3P) + \omega \rightarrow ... 2p^2\uparrow \ (^4P) + \varepsilon p\downarrow \ (^3D^o)$$

The partial cross sections for every photodetachment channel have been obtained in different approximations, which provides a possibility of evaluating the role of one or the other of the many-body correlations and their effect on the cross section behaviour (see for details the cited paper).

The calculations within the SP RPAE have shown that the $^3D^o$ partial contribution is primarily responsible for a sharp increase of cross section just after the $B^*(...2s\uparrow 2p^2\uparrow\ {}^4P)$ threshold (it resulted in a larger peak value of SP RPAE photoabsorption curve in comparison with the experiment). Within the SPHF frame the partial cross sections for the $2s\downarrow \rightarrow \varepsilon p\downarrow$ transition into the ^3D final state reveals the strong maximum near the $2s\downarrow$ threshold (fig.8) which is attributed to the transition into the quasi-bound "$...2s\uparrow 2p^2\uparrow 2p\downarrow$" ($^3D$) state. When taking into account intershell correlations, the interaction of the $2s\downarrow \rightarrow$ "$2p\downarrow$" ^3D resonance with the $2p\uparrow \rightarrow \varepsilon d\uparrow$ (^3D) channel leads to the interference structure, with a deep minimum being observed in the $\sigma_{2p\uparrow}$ ($^3D^o$) partial cross section (fig.9). The major part of the oscillator strength of the resonance is transferred to this channel: the resonance peak is shifted and considerably reduced (the RPAE curves in fig.8). For the photodetachment from the inner $2s\downarrow$ subshell with a final total symmetry of $^3D^o$ the dynamic core relaxation has been considered (curve 4 in fig.8). Due to screening the $2s\downarrow \rightarrow \varepsilon p\downarrow$ (^3D) photodetachment resonance peak is shifted to a higher photon energy, decreases and broadens. Taking account of the dynamic polarization of a core by an electron detached into p-continuum produces the opposite effect (curve 5 in fig.8). We have

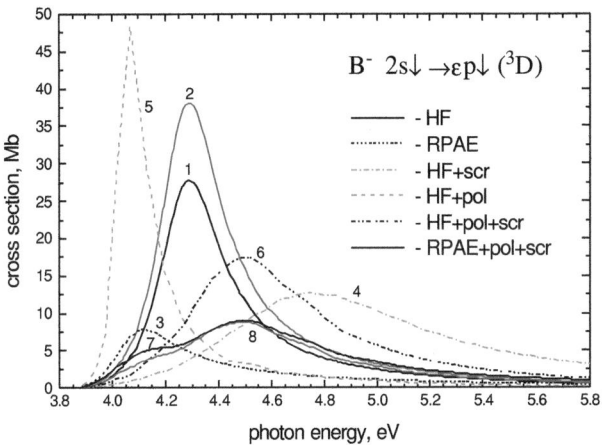

Figure 8. The $^3D^o$ partial cross section for 2s-photodetachment from the $...2s\downarrow 2\tilde{p}\uparrow^2$ (^3P) B^-: $B^-(^3P)+\omega \rightarrow B(^4P)+e^-(\varepsilon p)$ ($^3D^o$). The calculation are performed with the experimental ^4P threshold value. 1,2 - within the SPHF approximation in the r- and ∇- forms; 3- within the RPAE when interaction with $2p\uparrow \rightarrow \varepsilon d\uparrow$ (^3D) transition is taken into account. The curves 4 and 5 serve to illustrate the influence of the dynamic screening and atom-core polarization on a photodetachment process, respectively. 6 is the cross section in the single-particle approximation with allowance for the combined effects of the screening and polarization (3-6 present the r-form results); and finally 7,8 are the results with concurrent consideration of both these many-electron corrections and the RPAE correlations in the r- and ∇- forms.

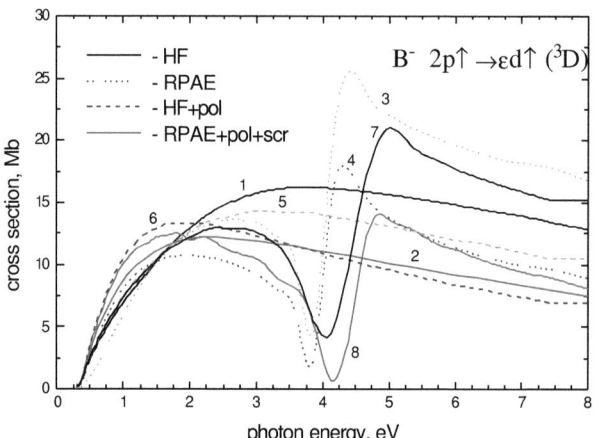

Figure 9. The $^3D^o$ partial cross section for 2p-photodetachment from the B^-: $B^-(^3P) + \omega \rightarrow B(^2P^o) + e^-(\varepsilon d)\,(^3D^o)$. The calculations are performed using the corrected within the DEM $2\tilde{p}$-wavefunction and the experimental $2P^o$ threshold value in a r- and ∇-forms of the dipole operator: 1,2 - within the HF approximation; 3,4 - with the RPAE correlations. 5,6 are the results of calculations which only include the dynamic polarization potential influence upon the detaching 2p-electron. The curves 7,8 depict the 2p-partial cross section obtained within the new method (with account of the interaction between the $2p\uparrow \rightarrow \varepsilon d \uparrow (^3D)$ and $2s\downarrow \rightarrow \varepsilon p \downarrow (^3D)$ transitions corrected for the dynamic polarization and screening effects).

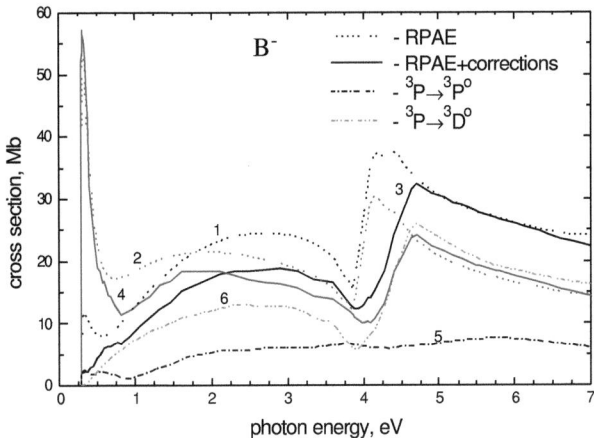

Figure 10. The total photodetachment cross section for B^- $...2s\downarrow 2\tilde{p}\uparrow^2$ (^3P) (in r- and ∇-forms): 1,2 - within the RPAE; 3,4 - within the method reported. The partial contributions in the r-form arising from the $^3P\rightarrow^3P^o$ (5) and $^3P\rightarrow^3D^o$ (6) total system symmetry transitions are also presented.

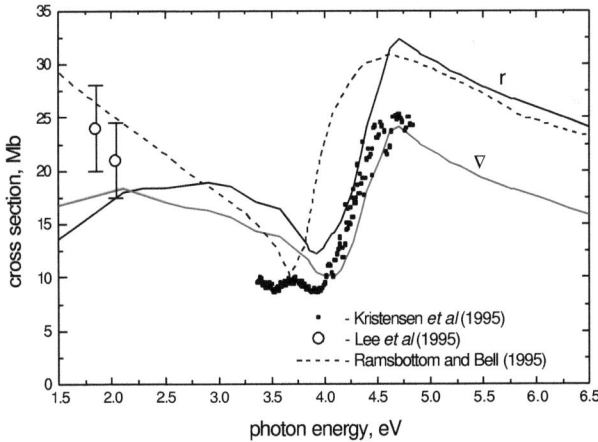

Figure 11. Photoabsorption in B: experiment (Kristensen et al. 1995, Lee et al. 1995) and theory (Ramsbottom and Bell 1995 and the present results).

considered the joint action of these factors (curve 6 in fig.8) and then the intershell interaction has been included (curves 6 and 7 in fig.8). The resonance position and width obtained ($\omega_{res} = 4.35 \pm 0.07\ eV$, $\Gamma = 0.82\ eV$) are close to the values determined by fitting the experimental photoabsorption curve to the modified Wigner threshold law (Peterson *et al* 1985): $\omega_{res} = 4.31\ eV$, $\Gamma = 1.16\ eV$ (Kristensen *et al* 1995). The "...$2s\uparrow 2p^2 \uparrow 2p\downarrow$" (^{3}D) shape resonance parameters estimated from the R-matrix partial cross section (Ramsbottom and Bell 1995) are $\omega_{res} = 4.22\ eV$, $\Gamma = 1.09\ eV$. Noteworthy also is an excellent agreement between the length and velocity forms of the resonance profile brought about by complete inclusion of collective effects. Owing to the improvement on the resonance parameters, the interference picture in the $2p\uparrow \to \varepsilon d\uparrow$ (^{3}D$^\circ$) partial cross section changes, the profile being smoothed and shifted (curves 7,8 in fig.9). It results in flattening of a peak in the total cross section (fig.10) compared to SP PRAE curves. In fig.10 we have also resolved the contributions to the total cross section arising from the transitions of ^{3}D$^\circ$ and ^{3}P$^\circ$ final total symmetries. Fig.11 makes it possible to compare our results with the experimental B$^-$ photoabsorption data (Kristensen *et al* 1995) and R-matrix theoretical total photodetachment cross section calculations (Ramsbottom and Bell 1995). The accord between theory and observation is achieved when the collective effects are properly accounted. Our many-body theory method provides good agreement with the experimental points with the correct position of the maximum and dip in the photoabsorption curve.

The results are presented in details in [3] (references in Appendix II)

8. Autodetachment "$3p^5 3d^6 4s^2$" resonance in Cr$^-$ photoabsorption

We have also directed our attention to calculations of partial and total Cr$^-$ photodetachment cross sections in the vicinity of the inner $3p^6$ shell threshold. The resonance associated with the transition of 3p electron to vacant states in 3d subshell has been revealed.

It is well known that existence of giant autoionizing resonances is a main feature in photoabsorption spectra of atoms with half-filled nd subshells. In particular, these have been found for photoabsorption from Cr and Mn neutral atoms (Amusia 1990) where the resonances appear due to interference of discrete $3p^63d^5{\rightarrow}3p^53d^6$ transitions of large oscillator strength and transitions into continua. The electron structure of the Cr⁻ ground state is the same as for neutral Mn atom: ...$3p^3{\uparrow}3p^3{\downarrow}3d^5{\uparrow}4s^1{\uparrow}4s^1{\downarrow}$ (^{6}S). It is known that there is the strong resonance reaching the maximum about 50 Mb with the width about 1.3 eV in Mn photoionization spectrum. The main contribution to the forming of this autoionizing maximum is introduced by the interference between the transition with large oscillator strength and the transition into continuum $3p^3{\uparrow}3p^3{\downarrow}3d^5{\uparrow}$ (^{6}S) + $\omega \rightarrow 3p^3{\uparrow}3p^2{\downarrow}3d^5{\uparrow}3d^1{\downarrow}$ (^{6}P) $\rightarrow 3p^3{\uparrow}3p^3{\downarrow}3d^4{\uparrow}\varepsilon f{\uparrow}$ (^{6}P) (Amusia et al 1983). Corresponding structure can be expected in photodetachment from Cr⁻ negative ion at photon energies just above the 3p threshold.

The refined wavefunctions (see Chapter 5) were used in the SP RPAE calculations. As expected, the quasi-bound "$3p^53d^64s^2$" state reveals itself as a strong resonance in the unperturbed $\varepsilon d{\downarrow}$ continuum in the immediate vicinity of the $3p{\downarrow}$ threshold (fig.12). The static relaxation of atomic neutral core was included by choosing photoelectron wavefunctions. The account of rearrangement leads to significant changes in the energy dependence of the partial photodetachment cross sections, it is especially true for the $3p{\downarrow} \rightarrow \varepsilon d{\downarrow}$ channel. Its maximum is shifted from the near-threshold region to the 47 eV and decreases (curves 2,3 in fig.12). Taking account of the intershell interactions combined with rearrangement effects leads to the equality of the cross sections in both of forms and results in further shifting of the peak value to the $3p{\downarrow}$ threshold and decreasing till 13 Mb (curve 4 in fig.12). The influence of the dynamic polarization was also considered for the partial $3p{\downarrow}\rightarrow \varepsilon d{\downarrow}$ cross within the method described. The result of calculations with account of the SP RPAE interactions and static rearrangement and dynamic polarization effects is presented by curve 5 in fig.12. The additional maximum in the partial $3p{\downarrow}\rightarrow \varepsilon d{\downarrow}$ cross section at

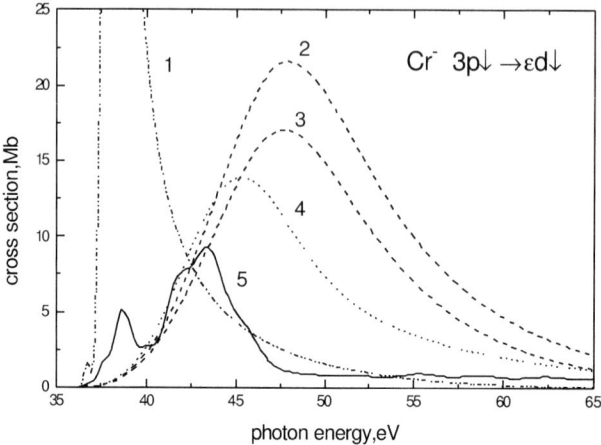

Figure 12. The partial $3p\downarrow \to \varepsilon d\downarrow$ photodetachment cross section for Cr^-: 1 - the SPHF calculations with corrected ground state wave functions (r-form); 2,3 - the SPHF calculations with consideration for the static rearrangement effects (r- and v- forms, respectively); 4 - the SP RPAE results with the static rearrangement taken into account (r-form); 5 - the results with the simultaneous inclusion of the RPAE corrections, the polarization (both for ground and excited states within the DEM) and the static rearrangement.

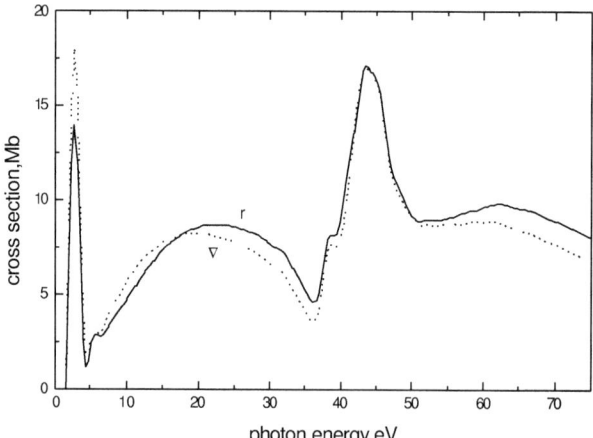

Figure 13. The total photoabsorption cross section for Cr^-.

around 38 eV appears due to opening inelastic scattering channel of the $4s\uparrow$ electron dipole excitations. We have analyzed the influence of interaction between strong $3p\downarrow\rightarrow\varepsilon d\downarrow$ channel and other possible transitions from 3p, 3d and 4s states. These correlations result in appearing interference structures corresponding to the position of the maximum in the $3p\downarrow\rightarrow\varepsilon d\downarrow$ photodetachment channel in the partial cross sections. The latter effect is most pronounced for the $3d\uparrow\rightarrow\varepsilon f\uparrow$ channel cross section

and also makes itself evident in the behaviour of the $3d\uparrow$ photoelectron angular asymmetry parameter. The influence of the $3p\downarrow\rightarrow\varepsilon d\downarrow$ resonance on dipole transitions from the outer 4s subshells is rather small resulting in small maxima in the partial cross sections. The total photodetachment cross section shown in the fig.13 reveals a complicated resonance structure to which a main contribution is made by the $3d\uparrow\rightarrow\varepsilon f\uparrow$ and $3p\downarrow\rightarrow\varepsilon d\downarrow$ transitions.

The report on the evidence of giant "$3p^5 3d^6 4s^2$" resonance in Cr^- photodedachment is in presented in Ivanov *et al.* 1997 ([4] reference in Appendix II).

9. Shape resonance in C⁻ outer-shell photodetachment

Before proceeding further, notice that although the elaboration of the new many-body theory method RPAE&DEM outlined above (section 7) was initiated in connection with working on the B⁻ photodetachment problem, this ion falls far short of being a good object in order to test the validity of the approach since it has an unfilled spin-polarized subshell. This made us extend the capabilities of the method. On the contrary, a series of the negative ions with half-filled p-subshells - C⁻, Si⁻ and Ge⁻ - is a canonical subject. For them window-type resonances were predicted within the SP RPAE which are associated with interference between the transitions from an inner s-subshell to a quasi-bound state in an open p-subshell and transitions of p-electrons to the continuum (Amusia *et al* 1990, Gribakin *et al* 1992). The existence of the resonance in Si⁻ has been confirmed experimentally by Balling *et al* (1993). The situation appears to be more uncertain for photodetachment from C⁻. A recent experiment (Haeffler *et al* 1996) has not found any resonance structure in the C⁻ photodetachment within the 5.23 - 6.04 eV range of photon energies. The R-matrix calculations (Ramsbottom *et al* 1993) predicted strong Feshbach resonance near the C⁻ $2s\downarrow$ threshold. Le Dourneuf *et al* (1995) reported theoretical cross sections which reveal shape resonance about 1eV above the threshold, just after the experimental points. Experimental data for higher energy region are not available now. For this reason we have decided to take a fresh look at C⁻ photodetachment.

Within the SPHF the electronic structure of the ion ground state can be presented as: C⁻ $1s\uparrow 1s\downarrow 2s\uparrow 2s\downarrow 2p\uparrow^3$ (^4S).

The extra $2p\uparrow$-electron in the C⁻ negative ion can be bound within the SPHF. The electron affinity for C and the outermost orbital wavefunctions have been refined within the Spin-Polarized version of DEM. The calculated values are $E_{2p\uparrow}^{SPHF}$ =-2.13 eV, $E_{2p\uparrow}^{DEM}$ =-1.36 eV, the experimental electron affinity is 1.26 eV (Scheer *et al* 1998). The corrected DEM $2\tilde{p}\uparrow$ wavefunction will be used in our further calculations.

Photodetachment calculations were performed within the new many-body theory method RPAE&DEM, interchannel interaction within the RPAE and dynamic relaxation and polarization corrections being included simultaneously. These effects are shown to be very important. So a shape resonance in the C⁻ 2s partial cross section turns out to be smoothed and far removed from the threshold when dynamic relaxation (screening) is taken into account (fig.14). It results in disappearance of sharp interference structure in $2p\uparrow \rightarrow \varepsilon d\uparrow$ partial cross section. Interaction between the transition into "$2p\uparrow^3 2p\downarrow$" quasi-bound state and the εd continuum in the frame of the RPAE along leads to a dip in the $2p\uparrow \rightarrow \varepsilon d\uparrow$ cross section just after the $2s\downarrow$ threshold followed by a sharp rise with redistribution of resonance channel oscillator strength. When resonance peak is shifted and broaden due to the screening corrections its effect on detaching d-wave is reduced and window resonance is not formed (fig.15). On the other hand, the correlation interaction moves the "$2p\uparrow^3 2p\downarrow$" state closer to the $2s\downarrow$ threshold, the energy position being sensitive to account of the dynamic

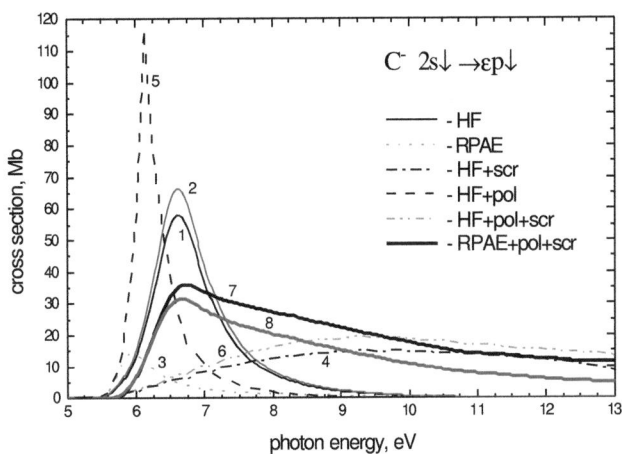

Figure 14. The $2s\downarrow \rightarrow \varepsilon p\downarrow$ partial cross section for photodetachment from the C^-. The calculation are performed with the experimental $2s\downarrow$ threshold value. 1,2 - within the SPHF approximation in the r- and ∇- forms; 3- within the RPAE when interaction with $2p\uparrow \rightarrow \varepsilon d\uparrow$ and $2p\uparrow \rightarrow \varepsilon s\uparrow$ transition is taken into account. The curves 4 and 5 serve to illustrate the influence of the dynamic screening and atom-core polarization on a photodetachment process, respectively. 6 is the cross section in the single-particle approximation with allowance for the combined effects of the screening and polarization (3-6 present the r-form results); and finally 7,8 are the results with concurrent consideration of both these many-electron corrections and the RPAE correlations in the r- and ∇- forms.

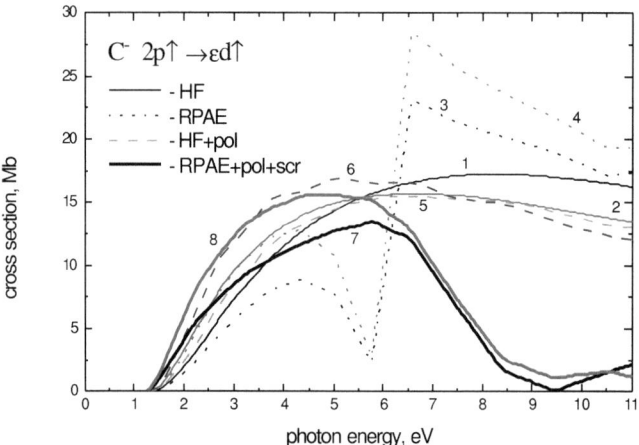

Figure 15. The $2p\!\uparrow \to \varepsilon d\!\uparrow$ partial cross section for photodetachment from the C. The calculations are performed using the corrected within the DEM $2\tilde{p}$-wavefunction and the experimental electron affinity value in a r- and V-forms of the dipole operator: 1,2 - within the HF approximation; 3,4 - with the RPAE correlations. 5,6 are the results of calculations which only include the dynamic polarization potential influence upon the detaching 2p-electron. The curves 7,8 depict the 2p-partial cross section obtained within the new method (with account of the interaction between the $2p\!\uparrow \to \varepsilon d\!\uparrow$ and $2s\!\downarrow \to \varepsilon p\!\downarrow$, $2p\!\uparrow \to \varepsilon s\!\uparrow$ transitions corrected for the dynamic polarization and screening effects).

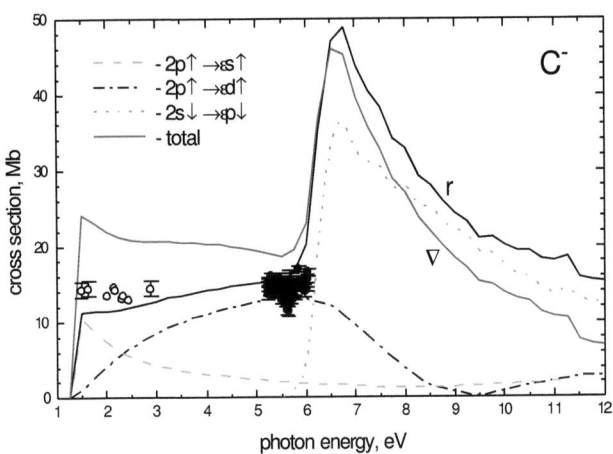

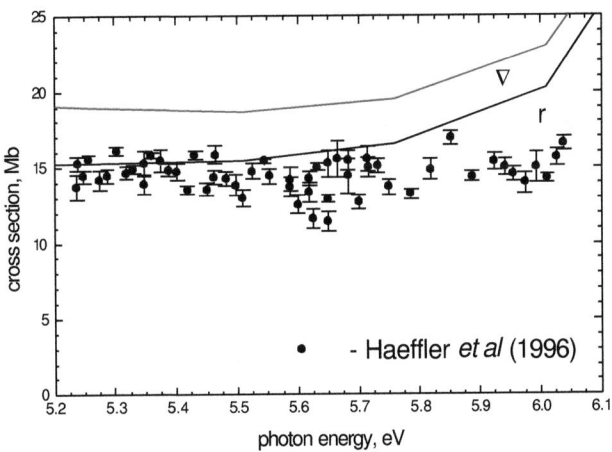

Figure 16. Photoabsorption in C⁻. The partial contribution (in r-form) and total cross section calculated with the interchannel interaction, dynamic polarization and relaxation taken into account. Experimental data: Seman and Branscomb (1962) (open circles) and Haeffler *et al* (1996).

polarization and screening. The shape resonance is found to peaks at the photon energy $\omega_{res} = 6.53$ eV ($E_{thr}(^5S) = 5.44$ eV). Our results reasonable good agree with the experimental finding (see fig.16).

So, as in the case of B⁻ photodetachment, the parameters for shape- and window-resonances in C⁻ near the 2s threshold appeared to be very sensitive to introduction of the corrections for dynamical relaxation and polarization – only their simultaneusly inclusion within RPAE&DEM brings the theoretical predictions in agreement with the experiment.

The results in details are presented in [5] (references in Appendix II).

10. Inner-shell C⁻ photodetachment:
Dramatic role of dynamic relaxation in collective response

In the beginning of XXI century a substantial progress has been achieved in describing and understanding the photodetachment processes (Ivanov and Kashenock 1999, Ivanov 1999, 2004, Andersen 2004, Pegg 2004, Gorczyca 2004, Kjeldsen 2006). However, experiments and subsequent theoretical interpretation were limited to outer-shells studies. New third-generation synchrotron light sources (ASTRID at University of Aarhus, IPB at the Advanced Light Source in Berkeley Lab) combined with advanced photoelectron spectroscopy and coincidence method have only recently made experiments possible on deep inner-shell photoionization of negative ions, with the first experimental investigations of inner-shell photodetachment being reported in 2001 (Kjeldsen *et al* 2001, Berrah *et al* 2001, 2002, Covington *et al* 2001). From the bibliography given in Appendix I on can see that in five years 2001-2005 the problem of inner-shell negative ion photodetachment turns to be a top-of-interest problem in atomic physics.

The inner-shell photodetachment experiments have been performed with the carbon negative ion C⁻ (Gibson *et al* 2003, Gorczyca 2004). Several theoretical calculations of the photodetachment cross sections within the frame of R-matrix approach have also been reported (Gibson *et al* 2003, Gorczyca 2004). However, the theoretical predictions are not in agreement with each other or with the latest experimental data (Gorczyca 2004). It was obviously, that the true nature of the observed strong near 1s-threshold resonance peak is far from being trivial.

A many-body theory investigation of the sharp near-threshold K-shell photodetachment resonance is the aim of this Chapter. We will focus on the inner-shell $1s$ photodetachment channel near the $(1s2s^2 2p^3\,^5S)$ threshold, which allows us to concentrate on one of the most important collective effects in the inner-shell photodetachment dynamics – core relaxation. We want to provide a deeper insight to the mechanisms of resonance formation and to the collective response of the spectator electrons in the presence of a core hole and an outgoing photoelectron.

In previous chapters we have applied the new many-body theory method RPAE&DEM to resonant outer-shell photodetachment: for He⁻ (Ivanov *et al* 1996a), B⁻ (Kashenock and Ivanov 1997), C⁻ (Kashenock and Ivanov 1998). Our results are in fairly good agreement with other available data for He⁻, B⁻, C⁻ (see the papers cited above and reference therein and subsequent works, *e.g.* Hi and Fischer 1996, Kim *et al* 1997, Ramsbottom and Bell 1999, Zhou *et al* 2004). In the present Chapter (it is based on the publication Ref. [6] in Appendix II) we re-examine our method with an even more sensitive probe for electron-correlations effects – the few-electron open-shell negative ion C⁻ and the resonance photodetachment from its deep inner shell.

10.1 DEM corrections to amplitudes of photoprocesses

For detailed description of the RPAE&DEM approach the reader is referred to Chapters 3 and 6 (refs. Ivanov *et al* 1996a, Kashenock and Ivanov 1997, 1998) in which the many-body theory method was developed for simultaneous inclusion of the dynamic polarization potential generated in the system "core + electron", the dynamic relaxation (screening) and the corrections within the scope of the RPAE. Here we repeat only main concepts of the theory with focus on the application of the Dyson equation method approach to describe inter-electron correlations. For the issue of this Chapter the inter-channel interaction within the RPAE is of minor importance.

One can see below that in the vicinity of the $1s$ inner-shell threshold the photodetachment of carbon negative ion C⁻ ($1s^2 2s^2 2p^3$ ^4S) can be considered as a one-channel process, the transition into p-continuum $1s \to \varepsilon p$. Nevertheless, the dynamical characteristics of the photoprocess definitely reflect the fact that the response of the atomic system to an external field is strongly collective.

The given above formalism – see equations (4), (6) and (8) - presents the DEM as the method which enables us to bind the outer electron in a negative ion and to correct the wavefunction and the single-particle energy of any electron, inner as well as outer, to take account for the influence of polarization interaction between a photoelectron and a core in the photodetachment process. $\Sigma_E(\vec{r}, \vec{r}')$ is equal to the self-energy part of the single-electron Green function. If the irreducible self-energy part Σ is chosen as (1) the equation (4) and (8) with the reducible self-energy part $\tilde{\Sigma}$ from (6) completely includes the polarization interaction and exchange (within the second order of perturbation theory) of the extra electron with the atom.

To incorporate the dynamic core-relaxation effects one should include the new type of diagrams comprising the mass operator involved as a kernel in the Dyson integral equation (2). We built up the screening interaction matrix $\langle \varepsilon'' | \hat{\Gamma}_{\varepsilon_0} | \varepsilon' \rangle$ with the following diagrams of the second order of the perturbation theory:

$$\tag{17}$$

The term "screening" needs clarification. Let us consider the amplitudes of effective interaction between particle and hole states, which obey the following equation (14) in more details:

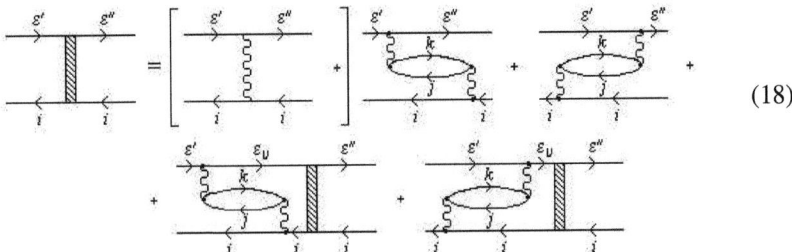

$$(18)$$

Here the block denotes the effective interaction between a particle and a hole $\langle\varepsilon''|\hat{\Gamma}_{\varepsilon_0}|\varepsilon'\rangle$. The first term (enclosed in brackets) in the right-hand side of equation (18) is presented in our calculations automatically since we use the "frozen core" approximation for definition of the photoelectron wavefunctions (see also equation (16) and the accompanying text). We find photoelectron wavefunctions in the "frozen" field of an ion with a hole (i), so that the allowance is made for the field of a hole. However, a hole field weakens due to screening - the interaction between an electron and a hole is not pure Coulomb one but is dynamically screened by other electrons of the core, with the outermost subshell (j) contribution being the most pronounced. This process is described by the infinite sequence of diagrams representing effective interaction amplitudes (18) as the sequence of the repeated diagrams (17).

Finally, to incorporate dynamical relaxation and polarization simultaneously within our approach we solve the following analytical integral Dyson equation

$$\langle\varepsilon''|\hat{\tilde{\Sigma}}_{\varepsilon_0}|\varepsilon'\rangle = \langle\varepsilon''|\hat{\Sigma}_{\varepsilon_0}|\varepsilon'\rangle + vp\int_{\varepsilon_v}\frac{\langle\varepsilon''|\hat{\tilde{\Sigma}}_{\varepsilon_0}|\varepsilon_v\rangle\langle\varepsilon_v|\hat{\Sigma}_{\varepsilon_0}|\varepsilon'\rangle}{\varepsilon_0-\varepsilon_v} + \langle\varepsilon''|\hat{\Gamma}_{\varepsilon_0}|\varepsilon'\rangle + vp\int_{\varepsilon_v}\frac{\langle\varepsilon''|\hat{\tilde{\Sigma}}_{\varepsilon_0}|\varepsilon_v\rangle\langle\varepsilon_v|\hat{\Gamma}_{\varepsilon_0}|\varepsilon'\rangle}{\varepsilon_0-\varepsilon_v} \quad (19)$$

It corresponds to the combination of the diagrammatical equations (6) and (18). After that the new effective amplitudes are calculated from (8) where the reducible self-energy part $\tilde{\Sigma}_\varepsilon$ comprises now the dynamical polarization and relaxation corrections.

10.2. Calculations of the ground and final states;influence of polarization on photoelectron states.

10.2.1 Ground and excited states within HF and RPAE

We start from the corrected within DEM configuration C⁻ $1s\uparrow 1s\downarrow 2s\uparrow 2s\downarrow 2\widetilde{2p}\uparrow^3$ (^4S) as it obtained in Chapter 9. The SPHF single-particle energies for subshells $1s\uparrow$ and $1s\downarrow$ are -298.63 eV and -297.32 eV, correspondingly. They differ from each other because within the spin-polarized approximations the exchange interaction between subshells with opposite spin is absent. This difference is of principal importance, since we consider "up" and "down" electrons as particles of the different types. Only from the $s\downarrow$ subshell the phototransition into the vacant "$2p\downarrow$" state is possible. Therefore, it is the transition $1s\downarrow\rightarrow\varepsilon p\downarrow$ that is a *resonant* inner-shell photodetachment channel similar to $2s\downarrow\rightarrow\varepsilon p\downarrow$ considered earlier (Chapter 8,9)

The wavefunctions of detached from the inner shell p-electron are calculated in "frozen-field" SPHF approximation with the unperturbed core $[1s\uparrow \underline{1s\downarrow} 2s\uparrow 2s\downarrow 2p\uparrow^3]$ (the hole is underlined) for $p\downarrow$-photoelectron and the core $[\underline{1s\uparrow} 1s\downarrow 2s\uparrow 2s\downarrow 2p\uparrow^3]$ for $p\uparrow$-photoelectron. The corresponding scattering phaseshifts for $\varepsilon p\downarrow$-continuum $\delta_l(\varepsilon)\rightarrow\pi$ as the energy ε goes to zero. According to Levinson's theorem this behavior indicates the presence of the bound $2p\downarrow$-state in the SPHF spectrum. The SPHF single-particle energy of this discrete excited state is $E_{2p\downarrow}^{SPHF}=-7.83$ eV. The calculated partial cross sections for the $1s\downarrow\rightarrow\varepsilon p\downarrow$ and $1s\uparrow\rightarrow\varepsilon p\uparrow$ photodetachment channels are presented in fig.17. The $1s\downarrow\rightarrow 2p\downarrow$ phototransition has rather strong oscillator strength $f_{1s\downarrow\rightarrow 2p\downarrow}^{SPHF}$ of 0.209 in the length (r-) representation of the dipole operator, or 0.212 in the velocity (∇-) form. So, the essential part of the oscillator strength falls on this resonant transition whereas the cross sections for the continuum are weak and the curves do not have a resonant character. We have considered inter-channel interaction between the two channels within the Spin-Polarized RPAE (SP RPAE) (the inter-channel correlations with outer subshells is negligible due to their energy farness). One can see in fig.17 that the SPHF and SP RPAE cross section curves are

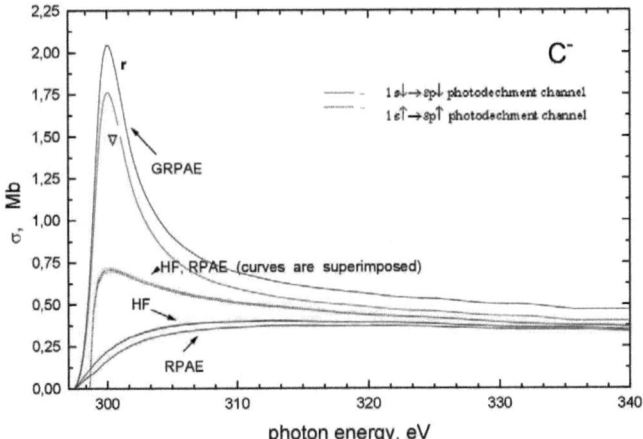

Figure 17. SP HF and RPAE partial cross sections for C⁻ $1s\downarrow\rightarrow\varepsilon p\downarrow$ (solid lines) and $1s\uparrow\rightarrow\varepsilon p\uparrow$ (dashed lines) photodechment channels. The $1s\downarrow\rightarrow\varepsilon p\downarrow$ -channel cross sections calculated within the simplest static relaxation approximation (GRPAE) are also presented in the **r**- and ∇ -forms (for the HF and RPAE results the difference between **r**- and ∇ -forms is negligible).

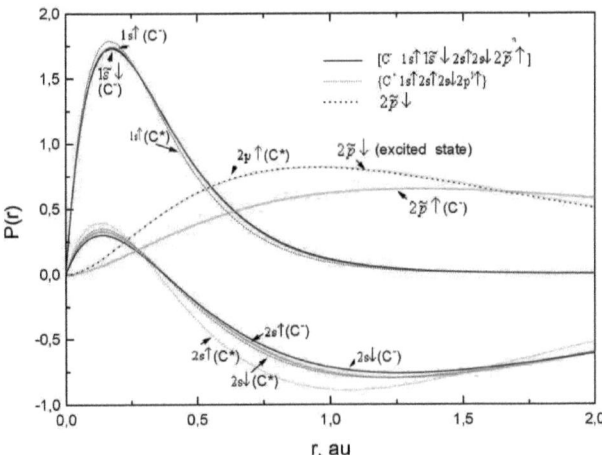

Figure 18. Radial wavefunctions for the SPHF configurations of the carbon negative ion ground state: [C⁻ $1s\uparrow1\tilde{s}\downarrow 2s\uparrow2s\downarrow1\tilde{p}\uparrow$] (solid lines), the excited carbon core {C* $1s\uparrow2s\uparrow2s\downarrow2p^3\uparrow$} (dashed lines) and a wavefunction of the discrete excited state of photoelectron $2p\downarrow$ (calculated in the "frozen" field of core [$1s\uparrow1\underline{s}\downarrow2s\uparrow2s\downarrow2p^3\uparrow$]) or $2\tilde{p}\uparrow$ (dotted line). The improved DEM wavefunctions marked with tilde are actually superimposed on SPHF ones.

practically indistinguishable over the whole photon energy range (except near the $1s\uparrow$-threshold region). Only less than 0.5% of $1s\downarrow \rightarrow 2p\downarrow$ phototransition oscillator strength is transferred through the RPAE inter-channel interaction ($f^{SPRPAE}_{1s\downarrow \rightarrow 2p\downarrow} = 0.200 /$ 0.203 for r- /∇- forms). So, we can consider the resonant $1s\downarrow$-photodetachment channel as not affected by other channels and concentrate on analysis of dynamical polarization and relaxation effects in this *one-channel* process.

In principle, the final stage of the photodetachment process "photoelectron + completely relaxed atomic core" can be described within the GRPAE approach. Then we find the new basis set of the SPHF wavefunctions for the neutral carbon atom ground state $\{C^* \; 1s\uparrow 2s\uparrow 2s\downarrow 2p^3\uparrow\}$ and calculate the SPHF continuum $p\downarrow$-wavefunctions in potential generated by it (let us denote these wavefunctions by $ep\downarrow$-states). The calculated cross sections are also presented in fig.17. We will see further that the pattern of resonant photodetachment is strongly smoothed in this approximation, as it might be expected for a deep inner shell from the above discussion (Section 2). Note that there is no a discrete transition in the spectrum within this approximation. The phase shifts of continuum $\{C^* \; 1s\uparrow 2s\uparrow 2s\downarrow 2p^3\uparrow\}ep\downarrow$ wavefunctions $\delta_1(e) \rightarrow 0$ as $e \rightarrow 0$, thus no bound states exist in $\{C^* \; 1s\uparrow 2s\uparrow 2s\downarrow 2p^3\uparrow\}$ potential. The phaseshift rises through the rather wide, ~ 7 eV, energy range near the threshold. However, this is rather weak evidence of a shape resonance in the continuum.

At this point it is reasonable to note that the existence and dynamics of the quasi-bound "$2p\downarrow$" state is a key point of our analysis of many-electron correlations in the system. The fact that this bound $2p\downarrow$ state exists within the accepted here zero-order independent-particle SPHF approximation makes the present analysis to radically differ from that for the "under-outermost" $2s\downarrow$ subshell photodetachment of C⁻ (Chapter 8,9). The experimental data on $1s$ photodetachment cross sections reported by Gibson *et al* (2003) are readily interpreted as an evidence of a shape resonance with the sharp rise at the threshold and decay tail above the threshold. The configuration which may be composed as "$[C^{-*} \; 1s\uparrow \underline{1s\downarrow} 2s\uparrow 2s\downarrow 2p^3\uparrow]2p\downarrow$" temporary negative ion or "resonance", can not be treated as a Feshbach resonance. Its total energy $E_{tot}(\text{"C}^-\text{*"})$=-675.49 eV is greater than that one for the parent configuration $\{\text{C}^*\ 1s\uparrow2s\uparrow2s\downarrow2p^3\uparrow\}$, $E_{tot}(\text{C}^*)$=-745.60 eV (although "C$^-$*" lays below the "frozen" configuration, $E_{tot}([1s\uparrow\underline{1s\downarrow}2s\uparrow2s\downarrow2p^3\uparrow])$=-617.40 eV). As typical for negative ions, there exists no real excited bound state for C$^-$. The quasi-bound state might be transformed into shape resonance as the result of collective correlations within the whole system "core + photoelectron". So we deal with subtle dynamics of the transient "$2p\downarrow$" state which is to be investigated.

10.2.2. DEM for $1s\downarrow$ state.

First of all the SPHF $1s\downarrow$-wavefunction has been refined within the DEM. The value $E_{1s\downarrow}^{DEM}$ =-281.84 eV is obtained as solution of (2) with the monopole, dipole and quadrupole electron excitations in the other subshells taken into account in calculating the irreducible self-energy part Σ from equation (1). As usual for inner shell energy corrections, the pure polarization diagrams in the self-energy part Σ (the first diagram in (1)) contribute very slightly. The main contribution is brought by the "time-reverse" diagrams

$$\begin{array}{ll} + & \text{exchange} \\ & \text{counterparts} \qquad\qquad (20) \\ & \text{(if exist)} \end{array}$$

The sum of corresponding monopole matrix elements $\left\langle 1s\downarrow\left|\Sigma_{E_{1s\downarrow}}^{(0),j}\right|1s\downarrow\right\rangle$ gives approximately the full DME correction to the $1s\downarrow$ energy value $\Delta E = E_{1s\downarrow}^{DEM} - E_{1s\downarrow}^{SPHF}$ =16.48 eV. This correction to the HF single-particle energy has the clear physical meaning: photodetachment from a deep inner shell results in adjusting of outermost electron orbitals due to reduced shielding of the nucleus, so we consider the core electron virtual excitations described by (20). (When taking to all orders these diagrams must bring the energy of hole into agreement with experiment.) Such rearrangement leads to reduction in the energy of ionic core and corresponding reduction of ionization threshold. The corrected single-particle $1\tilde{s}\downarrow$-energy is close to the theoretical value of $1s$ threshold proposed by Gibson *et al* (2003), $281.415\ eV$. The DEM $1\tilde{s}\downarrow$ wavefunction (although its corrected radial part is not changed radically) and energy $E_{1\tilde{s}\downarrow}^{DEM} = -281.836\ eV$ are used in the further calculations.

10.2.3. DEM for $2p\downarrow$ photoelectron state - influence of polarization

Outgoing photoelectron wavefunctions are also strongly affected by correlation interaction with core electrons and strongly "feel" the relaxation dynamics of the core. Fig.18 presents the radial wavefunctions for electron orbitals of the refined SPHF-DEM configuration of the ionic ground state [C $1s\uparrow 1\tilde{s}\downarrow 2s\uparrow 2s\downarrow 2\tilde{p}^3\uparrow$] and the excited carbon atom SPHF configuration {C* $1s\uparrow 2s\uparrow 2s\downarrow 2p^3\uparrow$}. We present curves for all these wavefunctions in order to show how wavefunctions rearrange from the initial state [C⁻] to an adiabatic limit – the final state {C*}. One can see that the ionic orbitals, especially the valence one, undergo visible changes. Depicted also in the same fig.18 is the SPHF radial wavefunction of the discrete $2p\downarrow$ excited state, $[1s\uparrow \underline{1s\downarrow} 2s\uparrow 2s\downarrow 2p^3\uparrow]2p\downarrow$. It is interesting to note that $2p\downarrow$ electron cloud takes up position inside the ionic cloud. It is obvious that photoelectron in this position must be strongly affected by $2s$ and $2p\uparrow$ core electrons. To take into account this interaction, which might be treated as polarization one, we solve the Dyson equation (2) for $2p\downarrow$-electron state with consideration of the monopole and dipole excitations of $j = 2\tilde{p}\uparrow$, $2s\downarrow$, and $2s\uparrow$ subshells for the irreducible self-energy matrix Σ from (1). Here, the main contribution to Σ and, hence, to the energy correction arises from the diagonal self-energy part matrix element (21)

$$= \langle 2p\downarrow | \Sigma_{2\tilde{p}\downarrow}^{(0),2\tilde{p}\uparrow} | 2p\downarrow \rangle = -3.56\ eV.$$

$$(21)$$

The refined discrete exited state $2\tilde{p}\downarrow$ has the single-particle energy $E^{DEM}_{2\tilde{p}\uparrow} = -11.39\ eV$.

These DEM $2\tilde{p}\downarrow$ wavefunction and energy are used for calculation of the effective single-particle dipole matrix element $\langle 2\tilde{p}\downarrow|\hat{d}|1\tilde{s}\downarrow\rangle$, thereby we take into account polarization corrections for the outgoing photoelectron state. The polarization influence on continuum state wavefunctions has been estimated by solving of equations (5) and (6) with the same irreducible self-energy matrix Σ. It is found that polarization practically does not affect the photodetachment amplitudes for the continuum final states. The oscillator strength for transition between DEM-refined states is $f_{1\tilde{s}\downarrow\rightarrow2\tilde{p}\downarrow}=0.195\ /\ 0.227$ (r- /∇-forms). The oscillator strength for $1s\downarrow\rightarrow2p\downarrow$ phototransition is not changed substantially due to only refinement of the description of discrete $2\tilde{p}\downarrow$ states beside the HF approximation ($f_{1\tilde{s}\downarrow\rightarrow2p\downarrow} = 0.198\ /\ 0\ .224$). However, we show below that the polarization corrections are very important when they superpose on dynamical relaxation corrections. They, as usual (see e.g. the cases of $2s\downarrow$-photodetachment for B$^-$ or C$^-$, Chapters 8,9), counteract the relaxation effects.

10.3. Dynamical relaxation influence

10.3.1. DEM for relaxation effects in 1s C$^-$ photodetachment

The decisive role in forming of the resonance structure in the $1s$ near-threshold photodetachment proved to belong to dynamical relaxation correlations. We introduce them in the description of photoelectron propagator by solving the Dyson equation (19) for reducible self-energy part $\tilde{\Sigma}$ with effective screening interaction $\tilde{\Gamma}$ included. Analytically, it means to solve equation (9), where the polarization part of correlations Σ is taken by substitution of DEM wavefunction for $2\tilde{p}\downarrow$ state and Γ is taken as equation (17) with $j=2\tilde{p}\uparrow$. Then the effective dipole amplitudes are found from (8) and diagrammatically looks like as:

$$\langle 2\tilde{p}\downarrow,\varepsilon p\downarrow|\hat{D}_{\varepsilon}|1\tilde{s}\downarrow\rangle = \qquad\qquad\qquad\qquad\qquad\qquad\qquad\qquad\qquad\qquad (22)$$

51

The oscillator strength and cross sections obtained with these amplitudes present the net result of consideration for dynamical polarization and relaxation effects on the photoelectron initial, virtual intermediate and final states. It includes the correlation response of the atomic core on the photodetachment process. The calculated cross sections are displayed in fig.19. The oscillator strength for the phototransition into the $2\tilde{p}\downarrow$ state with effective screening interactions (dynamical relaxation) taken into account is $f_{1\tilde{s}\downarrow\rightarrow2\tilde{p}\downarrow}= 0.0037 / 0.0046$ (**r**- /∇ -forms), *i.e.* two orders of magnitude less than that without relaxation corrections. Observable changes in the cross sections with displacing oscillator strength to higher photon energies due to inclusion of core relaxation are typical for inner-shell photodetachment. The outgoing photoelectron receives a boost in kinetic energy as the atomic core becomes less attractive due to screening of the hole field. However, in our case of the deep inner-shell photodetachment the pattern changes radically. The majority of the oscillator strength of the discrete transition is transferred into continuum where the strong resonance is formed near the threshold after including the collective effects, with the relaxation effect being the most pronounced and critical.

10.3.2. Threshold energy, resonance parameters

The results obtained are in perfect agreement with the most recent experimental data (the points in fig.19) on the absolute photodetachment cross sections in the energy range of interest (Gorczyca 2004). The theoretical curves are shifted in photon energy by -0.88 *eV* so that the peak position coincides with the experiment (281.814 *eV*). We justify this correction to the threshold position in the following way. According to the Koopmans' theorem the $1s\downarrow$ threshold value (a rather crude approximation, especially for an inner shell) is equal to $- E_{1\tilde{s}\downarrow}^{HF} = 297.320$ *eV*. The difference between the SPHF total energy of the carbon excited atom-core and the total energy of carbon negative ion also gives the estimation for the threshold energy (adiabatic treatment). One obtains $E_{tot}(C^*)$ - $E_{tot}(C^-)$ = -745.599-(-1026.100)=280.501 *eV*. With many electron correlations taken into account within the DEM the *ab initio* threshold value

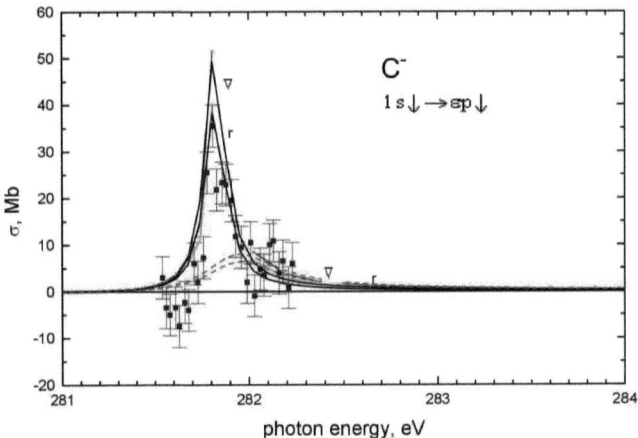

Figure 19. C⁻ inner-shell photodetachment cross section. The dynamical relaxation and polarization corrections are included. The results of calculations (in **r**- and ∇ - gauge, thin crossed lines) and the best-fit profiles (solid lines) are displayed. The experimental points are from Gorczyca (2004). The dashed lines present cross section with only relaxation effect considered.

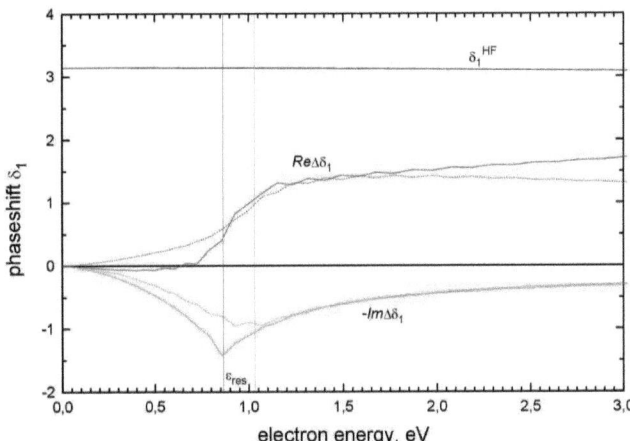

Figure 20. *p*-wave scattering phaseshifts within the HF "frozen" core approximation and with dynamical correlation interaction included (corrections denoted as $\Delta\delta_1$). The dotted lines correspond to taking into account for only relaxation effects.

is assumed to be $-E_{1\bar{s}\downarrow}^{DEM} = 281.836\ eV$. Our shift in energy is just due to corrections of higher orders that we neglect within our approach. The $1s$ threshold is not known experimentally. Pederson *et al* (1985) have derived the parametric formula for the modified Wigner threshold low in the case of resonance nearness, Feshbach as well as shape resonance. The fitting to our shifted cross sections to (9) gives the expected experimental threshold value of $E_{thr} = 280.957\ eV$.

The resonance parameters found from (9) are the resonance energy of E_{res} $= 281.814\ eV$ and width of $\Gamma = 0.122\ eV$ for the both "length" and "velocity" gauges. In fig.19 we also display the fitting curves for our results which well describe not only the behaviour of the cross section in the immediate vicinity of $1s$ threshold but the resonance peak itself with the exception of high-energy tail which we cut at 285 eV in the fitting. We choose the cut-off energy in such way that the oscillator strength under the resonance peak corresponds to the oscillator strength for discrete transition, $f_{1s\downarrow \to 2p\downarrow}^{SPHF}$, without corrections. Note, the resonance peak itself is very narrow and the long-energy tail gives very significant contribution to the total oscillator strength (see also fig.17). Really, we have to consider the photon energy range up to 500 eV, then $f_{1s\downarrow}$ is close to 1 as it must be according to the Thomas-Reiche-Kuhn (TRK) sum rule (it is important for saturation of the sums in equations (5) and (9)). The oscillator strength integrated over 500 eV range under the fitting curve is substantially less than for the calculated curves.

10.3.3. *Photoelectron scattering phaseshift*

We present also the scattering p-wave phaseshift (fig.20) calculated within the "frozen" core SPHF approximations δ_1^{HF} and corrections to it within the DEM $\Delta\delta_1$, when dynamical polarization and relaxation are included. The correction $\Delta\delta_1$ is defined by the diagonal matrix element of the reducible self-energy part (11). The background HF phaseshift δ_1^{HF} varies slowly rising from π, whereas the real part of the correction demonstrates resonance behaviour – it rapidly rises over a narrow energy region. The standard Breit-Wigner analysis for an isolated resonance is scarcely applicable in our case of the resonance whose width is comparable with its energy. Moreover, we have the situation of mixing the resonance types, in fact, quasi-bound state ("Feshbach") passes into continuum ("shape"). However, the energy position relative to the threshold ε_{res} =0.857 eV corresponding to the maximum at E_{res} =281.814 eV is well defined by elastic scattering phaseshift step jump. The imaginary part of the correction (an inelastic scattering characteristic) has the extremum corresponding to the resonance position that is also a "trace" of its discrete state origin.

It seems to be constructive to compare the results with that obtained without consideration for dynamical polarization influence on photoelectron state (Figures 19, 20, dashed curves). When dynamical screening is accounted, the effective oscillator strength for $1\tilde{s}\downarrow \rightarrow 2p$ transition is changed slightly less ($F_{1\tilde{s}\downarrow \rightarrow 2p\downarrow}$= 0.0042 / 0.0051 for r- /∇-forms) than that for the $2\tilde{p}\downarrow$ state. However, the resonance position and width, its peak value are changed drastically compared with the case of inclusion polarization and relaxation simultaneously. The less part of the oscillator strength is transferred into continuum and the resonance is crudely smoothed. So, as it was stated above, all the correlations are substantial for proper description of complex resonance structure in such a strongly correlated system.

10.3.4. Photodetachment cross section

Fig.21 reproduces the comparison between the present results, the experimental data and the theoretical cross sections calculated within the R-matrix approach. The latter methodology is referred in the cited paper (Gorczyca 2004) as the method R-matrix-I (cross sections are shifted by +0.05 eV) relative to their theoretical threshold 281.415 eV), the method II (cross sections are shifted by -0.98 eV) and the method (I + extra CI). In that calculations the authors (Gibson *et al* 2003, Gorczyca 2004) remark the high sensitivity of the results to the details of accounting for correlations due to mixed nature of the resonance and identify the resonance as of the "shape" nature rather than Feshbach one. Our study on the subject is complementary and convincing

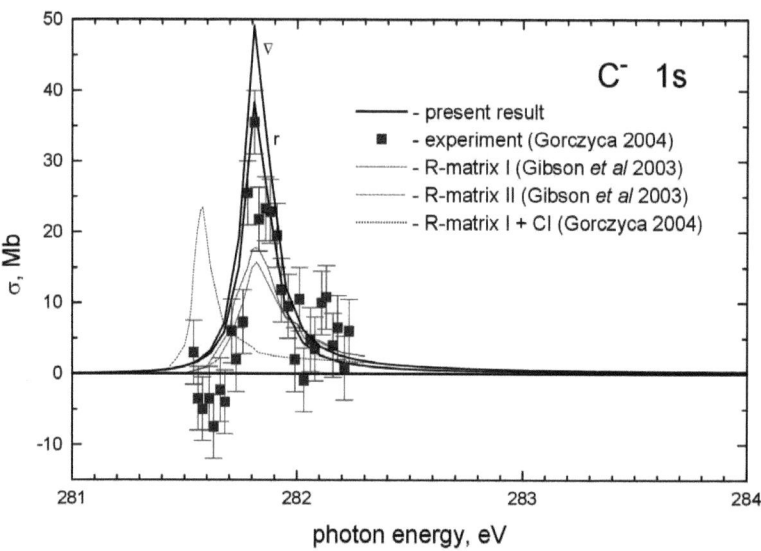

Figure 21. C⁻ inner-shell photodetachment cross section: comparison of the present result with the experimental data and theoretical R-matrix calculations (Gibson *et al* 2003, Gorczyca 2004). The precise resonance parameters from the higher resolution spectra (Walter *et al*. 2006) - the resonance energy $E_{res} = 281.74(10)$, the width $\Gamma=0.11(4)\ eV$, the threshold energy $E_{thr} = 281.64(12)$.

evidence of the fact that the formation of the resonance structure in the inner-shell C^- photodetachment is a process with subtle complex dynamics inherent. Being classified as a shape resonance, the type called also "one-channel resonance", it is actually strongly affected by the collective response of the whole system.

Shortly after the present results were published (Kashenock and Ivanov 2006) the improved experimental data have been obtained for K-shell photodetachment from C^- (Walter *et al* 2006). The precise measured resonance parameters - the resonance energy E_{res} = 281.74(10) and the width Γ=0.11(4) eV - and the threshold energy E_{thr} = 281.64(12) turned out to be in very good agreement with the theoretical ones and the authors also come to the conclusion that "comparison of these experimental results with advanced theoretical calculations demonstrate the importance of multielectron collective interactions in the behavior of negative ion".

To finalize, in this chapter we have considered strongly correlated system, the few-electron open-shell negative ion C^-, a perfect representative example for the many-body theory study of inner-shell photodetachment processes. On the one hand, we deal with a direct one-channel process, phototransition of a deep core electron into p-continuum, which is well described by experimental data on the near-threshold resonance peak in photodetachment cross section. On the other hand, close look reveals complex nature of the resonance under investigation. The whole atomic system turns out to be involved in resonant photodetachment process, core electrons and outgoing photoelectron being strongly coupled by correlation interaction. It is shown in the present work that only comprehensive inclusion of dynamical collective effects on resonance formation in photodetachment gives the adequate description of the process. The dynamical relaxation of the carbon core is responsible for transformation of a quasi-bound 2p state into the strong narrow shape resonance in the $1s$ C^- photodetachment continuum. However, the resonance maintains the mixed nature, of Feshbach-type origin, affected also by the dynamical polarization correlations. Being a perfect test-bed for the many-body theory method applied, the challenge has demonstrated the approach ability of accurate inclusion of collective effects in its complex interplay.

11. Si⁻ inner-shell photodetachment

In the concluding Chapter the very new results on inner-shell photodetachment for Si⁻ $(1s^22s^23p^63s^23p^3$ $^4S^o)$ negative ion will be presented. One can expected that the possibility of a photoexitation to the ion state Si⁻* $(1s^22s2p^63s^23p^4$ $^4P)$ reveals itself as a resonance structure in photodetachment cross sections in the energy range of the 2s and 2p inner shells thresholds similar to the 1s inner-shell photodetachment from C⁻ (Chapter 10). The system is more complex compared to C⁻ since we need to consider the partial cross sections for 4 close spin-polarized inner subshells (6 phototransitions):

$$\text{Si}^- \ldots 2p^3\uparrow\downarrow\ldots 3p^3\uparrow\ (^4S) + \omega \to \text{Si} \ldots 2p^2\uparrow\downarrow\ldots 3p^3\uparrow\ (^3P) + \varepsilon s\uparrow\downarrow; \varepsilon d\uparrow\downarrow\ ;$$
$$\text{Si}^- \ldots 2s\uparrow 2s\downarrow\ldots 3p^3\uparrow\ (^4S) + \omega \to \text{Si} \ldots 2s\downarrow\ldots 3p^3\uparrow\ (^4S) + \varepsilon p\uparrow;$$
$$\text{Si}^- \ldots 2s\uparrow 2s\downarrow\ldots 3p^3\uparrow\ (^4S) + \omega \to \text{Si} \ldots 2s\uparrow\ldots 3p^3\uparrow\ (^4S) + n,\varepsilon p\downarrow.$$

For the last phototransition we have emphasized the existence of photoexitation to the $3p\downarrow$ (n=3) discrete state in half-filled outer p-shell. This channel is expected to be a "resonance channel" as we have seen in the case of "$1s2s^22p^4$" resonance in C⁻. However, the resonance channel for inner-shell photodetachment is open at the $2s\downarrow$ threshold ($E_{2s\downarrow}^{SPHF}$=-11.784 Ry) in the vicinity of the thresholds of the others inner-shells photodetachment channels: $E_{2s\uparrow}^{SPHF}$=-11.806 Ry, $E_{2p\uparrow}^{SPHF}$=-8.020 Ry, $E_{2p\downarrow}^{SPHF}$=-7.973 Ry. So the RPAE correlations become important.

In this Chapter we will perform the analysis of the collective response of the ionic many-electron system Si⁻ on an electromagnetic field in the different levels of approximation: the "frozen-field" RPAE, the static relaxation approximation (GRPAE) and also within the DEM&RPAE approach when the dynamic relaxation and polarization will be included simultaneously with the RPAE corrections.

Fig.22 presents the results of calculation in the "frozen-field" spin-polarized RPAE (Amusia 1990) when the photoelectron wavefunctions are obtained with the same SPHF Hamiltonian as for the initial Si⁻ $1s\uparrow 1s\downarrow 2s\uparrow 2s\downarrow 2p^3\uparrow 2p^3\downarrow 3s\uparrow 3s\downarrow 3p^3\uparrow$ configuration but with a hole in the corresponding subshell $2s\uparrow$, $2s\downarrow$, $2p\uparrow$ or $2p\downarrow$. In this approximation the "3p↓" resonance state is a real bound state of an electron in the field of ion with a hole 2s↓ (neutral system). The oscillator strength $f_{2s\downarrow\to3p\downarrow}^{SPHF}$ = 0.0536 corresponding to phototransition 2s↓→ 3p↓ reveals itself only as the interference, Fano profiles in the partial cross sections of photodetachment from 2p subshels at the photon energies corresponding to the energy of transition into 3p↓ state ($E_{3p\downarrow}^{SPHF}$=-0.31256 Ry). Due to the interchannel interaction the part of oscillator strength transferred to other channels and $f_{2s\downarrow\to3p\downarrow}^{RPAE}$ is of 0.02925 in the length (r-) representation of the dipole operator, or 0.03329 in the velocity (∇-) form. The partial cross section for 2s↓→ εp↓ channel gives only very weak background ≈ 0.2 Mb that even becomes less when the interactions with other channels is allowed for.

To consider the other limit of the approximation we use the Generalized RPAE (Amusia 1990) and calculate the photoelectron wavefunctions in the field of the completely rearranged spin-polarized Hartree-Fock configurations without a detaching electron (fig.23). As was mentioned in Chapter 2 this approach (and its relativistic analogue, the RRPAR, Radojević et al 1989, Kutzner 2004) includes a simplification that the core rearrangement due to a hole creation is taken into account instantly wherever the photon is absorbed and photoelectron starts to move. Until the recent time, without DEM&RAPE approach having in hand, we had only used this approach suggesting that the relaxation time is less than the electron escape time. However, within the static approximation the role of relaxation effects often turns out to be overestimated, the near-threshold photodetachment picture being distorted (Ivanov 1999). At the GRPAE level p-photoelectron cannot be bound into 3p↓ state in the SPHF field of the rearranged Si atom without 2s↓ electron. However, the p↓-wave phaseshift demonstrates typical for a shape resonance behavior: $\delta_{1\downarrow}^{GRPAE}(\varepsilon)$ sharply increases from the value π at zero photoelectron energy to 5.2 over the narrow photoelectron energy range (note, that in "frozen-field" approximation $\delta_{1\downarrow}^{HF}(\varepsilon) \to 2\pi$, $\varepsilon \to 0$ - see fig.26 below - that corresponds to an existence of two bound p↓ states). Correspondently, in the partial 2s↓→ εp↓ cross section appears

Si⁻. "FROZEN-FIELD" (RPAE). PARTIAL CROSS SECTIONS.

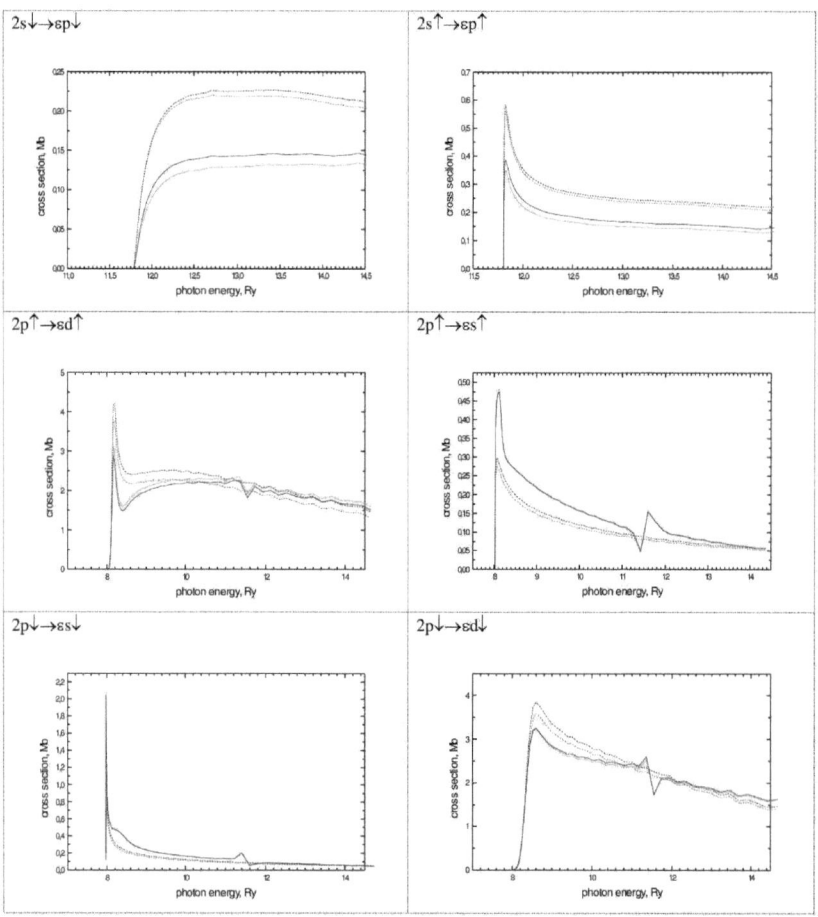

Figure 22. Partial cross sections for the inner 2s and 2p shells of Si⁻ calculated in the "frozen field" approximation: black (**r**-form) and red (∇-form) lines – in the one-particle SPHF approximation with the wavefunctions of photoelectron founded in the field of "frozen" core; blue (**r**-form) and green (∇-form) lines – with the account of the RPAE interactions between six channels.

Sī. STATIC RELAXATION (GRPAE). PARTIAL CROSS SECTIONS.

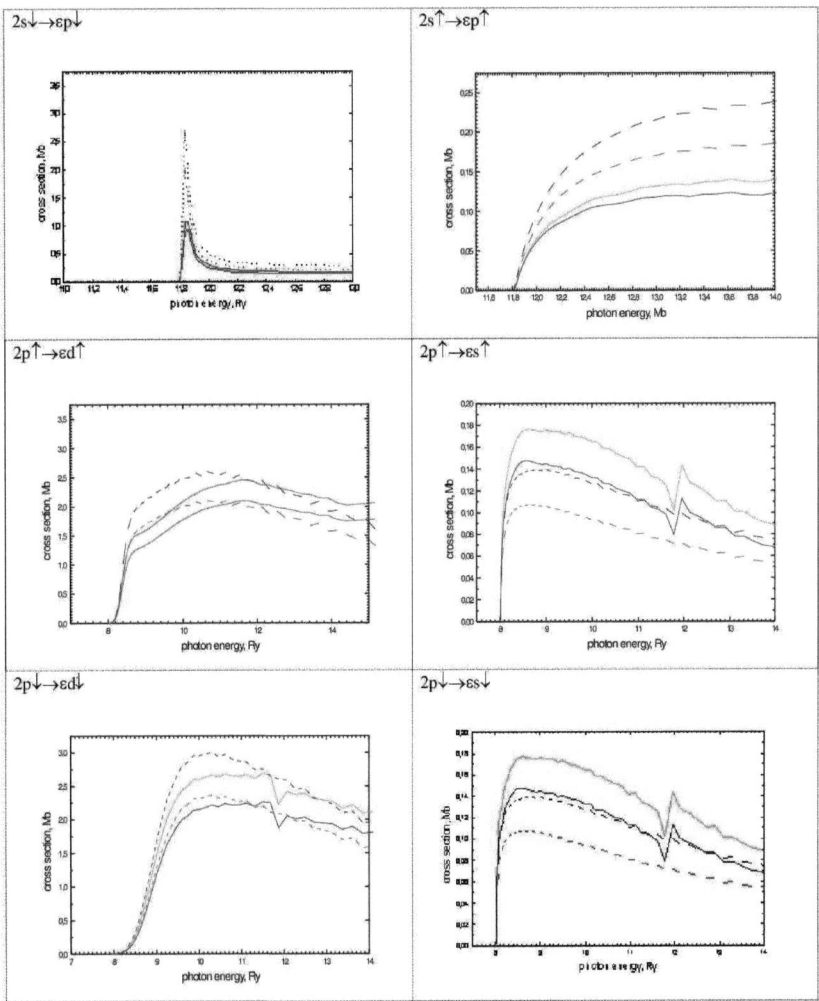

Figure 23. Partial cross sections for the inner 2s and 2p shells of Sī calculated in the static relaxation approximation: black (**r**-form) and red (∇-form) lines – in the one-particle SPHF approximation with the wavefunctions of photoelectron founded in the field of completely rearranged core; blue (**r**-form) and green (∇-form) lines– with account of the GRPAE interactions between six channels.

Si⁻. STATIC & "FROZEN-FIELD" APPROXIMATIONS. GRPAE *vs* RPAE.

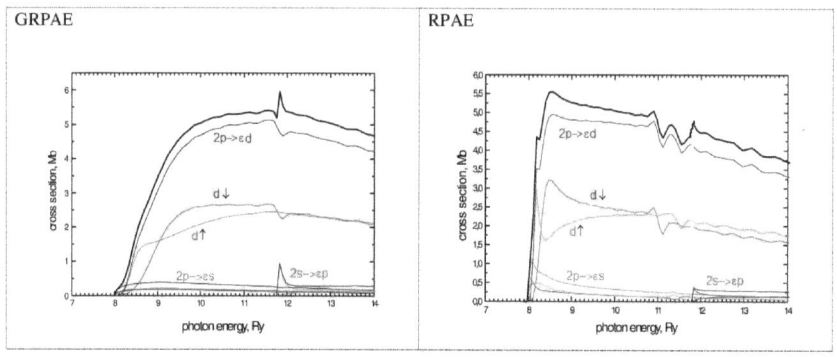

Figure 24. Total photodetachment cross section in "frozen field" and static relaxation approximations. The partial cross sections from figs. 22 and 23 and their sums are also depicted.

Figure caption (see below):

Figure 26. The results of calculations within the DEM&RPAE method – the dynamical relaxation, polarization and interchannel interaction are included.
Upper panel: The photoelectron angular asymmetry parameter for 2p shell Si⁻ in **r**-form (black) and ∇-form (red) of a dipole operator.
Middle panel: The scattering phaseshift for an $\varepsilon p\!\downarrow$-photoelectron. The phaseshift $\delta_1(\varepsilon)$ is divided into three components: the HF phaseshift $\delta_1^{HF}(\varepsilon)$ (black line) and an additional phaseshift $\Delta\delta_1(\varepsilon)$ due to the dynamical relaxation and polarization corrections accounted within the within DEM: $\mathcal{R}e\Delta\delta_1$ (red line) and $\mathfrak{Im}\Delta\delta_1$ (green line); dotted lines correspond to consideration only the critical for formation of the resonance relaxation effect without polarization.
Lower panel: The total photodetachment cross section for the inner 2s and 2p shells of Si⁻ calculated within the DEM&RPAE method. The partial cross sections from fig. 25 are also depicted.

resonant near-threshold peak (fig 23), however, with maximal value only of ≈ 2.5 Mb (1Mb when interchannel interaction is accounted). In partial cross sections for the 2p subshells due to the interchannel interaction the interference profiles arise just after the 2s↓ threshold. It is worth to note, that in the static relaxation approximation the weak shape resonance structure in 2s↑→ εp↑ shown in fig.22 ("frozen field") disappears – quasi bound "4p↑" not exists in this approximation of a strong relaxation. The total cross sections for Si⁻ in the region of the 2s,2p thresholds reveals not any prominent resonance structure in the GRPAE as well as in the "frozen-field" the approximation (fig.24).

The existence of the both limit "3p↓" states - as a bound state, or Feschbach resonance, in the "frozen-core" approximation and a quasi-bound state, shape resonance, in the static relaxation approximation - allow us to suppose that the real situation is subtler and could be close to that in the C⁻ inner-shell photodetachment. Due to strong electron correlation in many-electron system the resonance type could be considered as a mixed Feschbach-shape structure. To investigate the many-electron mechanism of forming the resonance near 3s↓ threshold in details we have used DEM&RPAE approach (Chapter 6 and 10.1). The one-particle approximation corresponds to that accepted in the "frozen-field" approximation. However, when dynamical relaxation and polarization are included even only for one, resonance 2s↓→ εp↓ channel the photodetachment pattern is changed dramatically. When we have found the photodetachment amplitudes for the resonance 2s↓→ εp↓ channel from the equation (8), where the reducible self-energy part $\tilde{\Sigma}$ consists of the dynamical relaxation and polarization parts, the partial cross section reveals a very strong, narrow near-threshold resonance (fig.25).

In the frame of the DEM&RPAE approach discribed in Chapters 3,6,10 we initially introduce the dynamical polarization correction by calculating self-energy part $\Sigma_E(\vec{r}, \vec{r}')$ (1) in the second order of the perturbation theory. With these corrections the DEM binding energy for "3p↓" resonance state is $E_{3p\downarrow}^{DEM} = -0.49678$ Ry. The corrected DEM $\widetilde{3p\downarrow}$ photoelectron wavefunction is used for the further calculation.

When the dynamic relaxation, the screened interaction of a photoelectron with the hole 2s↓ is taken into account by solving (8), (18), (19) with the screening diagrams (17) where i is 2s↓, the oscillator strength corresponding to 2s↓→ $\widetilde{3p}$↓ phototransition turn out to be practically fully transferred to εp↓ continuum. The oscillator strength $f_{2s\downarrow\rightarrow3\tilde{p}\downarrow}^{DEM}$ is equal 0.0046 / 0.0049 (**r**- /∇-forms) when the dynamical relaxation and polarization corrections are included, *i.e.* one order less than that in the one-particle approximation (and even two orders less without account of a polarization effect, due to relaxation only). We predict the strong resonance peak at the energy E_{res}=-11.82 Ry, ε_{res}= 0.04 Ry with resonance width of Γ=0.02 Ry. It should be noted here that in what follow we use the SPHF value for threshold energies, so for a comparison with possible experiments the resonance region in fig.26 should be shifted, since our DEM correction to threshold 2s↓ energy gives the more precise threshold value $E_{2s\downarrow}^{DEM}$=-11.408 Ry. The estimation for this value according to Koopman's theorem $E(\mathrm{Si}^*)_{tot}^{SPHF} - E(\mathrm{Si}^-)_{tot}^{SPHF}$ gives 11.112 Ry.

The middle panel of fig.26 shows the photoelectron phaseshift $\delta_1(\varepsilon)$: the p↓-wave phaseshift calculated in the one-particle SPHF approximation and also the corrections to it due to the dynamical relaxation and polarization within the DEM. One can see that the real part of correction (red line) demonstrates the characteristic behavior for a classical shape resonance. In the same time the reducible self-energy part and, correspondingly, the phaseshift have also an imaginary part (green line) that corresponds to a real bound state – a Feschbach resonance.

When the interchannel interaction is included within the DEM&RPAE the shape resonance peak stays practically unchanged, however it affects strongly on partial photodetachment from the nearest 2s↑ subshell (destructive interference) and in the less extent on the partial crosssections for 2p electrons photodetachment (fig. 25). One can see from fig.25 that the Fano profiles due to interchannel interaction give the clear "trace" of an origin bound state "3p↓", the Feschbach resonace localized before the 2s↓ threshold. Correspondingly, the interaction with the other channels affects the oscillator strength for 2s↓→ $\widetilde{3p}$↓ phototransition – $f_{2s\downarrow\rightarrow3\tilde{p}\downarrow}^{DEM\&RPAE} =$

Si⁻. RPAE&DEM. PARTIAL CROSS SECTIONS.

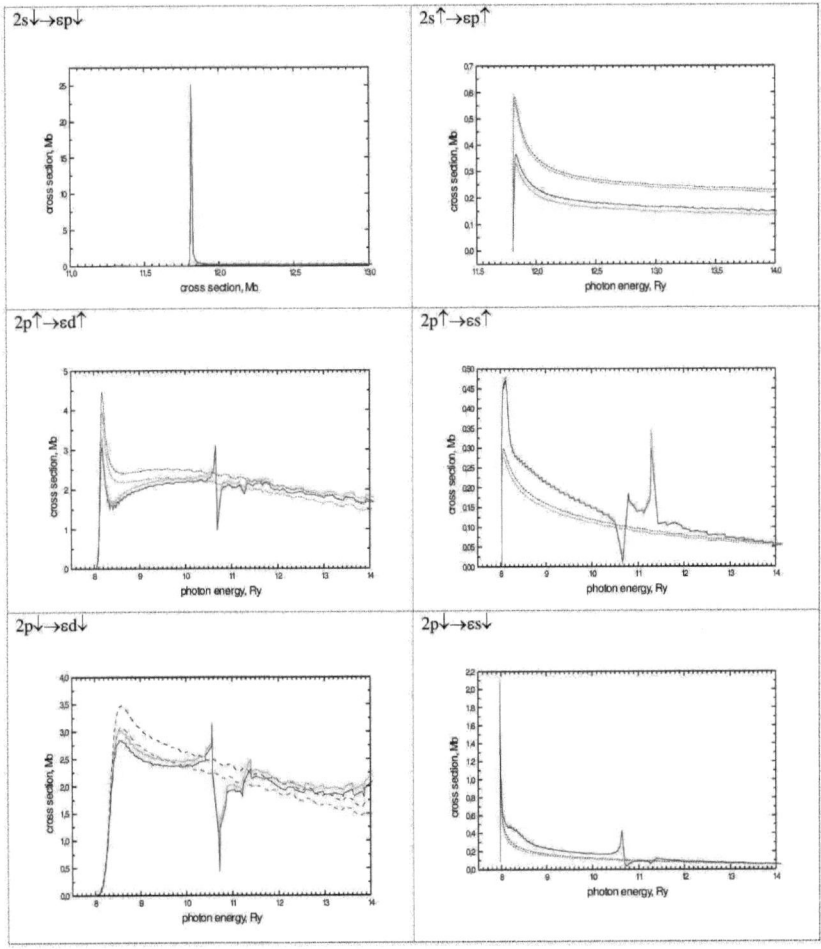

Figure 25. Partial cross sections for the inner 2s and 2p shells of Si⁻ calculated within the DEM&RPAE method: black (**r**-form) and red (∇-form) lines – in the one-particle SPHF approximation with the wavefunctions of photoelectron founded in the field of "frozen" core; blue (**r**-form) and green (∇-form) lines – with account of the dynamical relaxation, polarization and RPAE corrections.

Si⁻. RPAE&DEM.

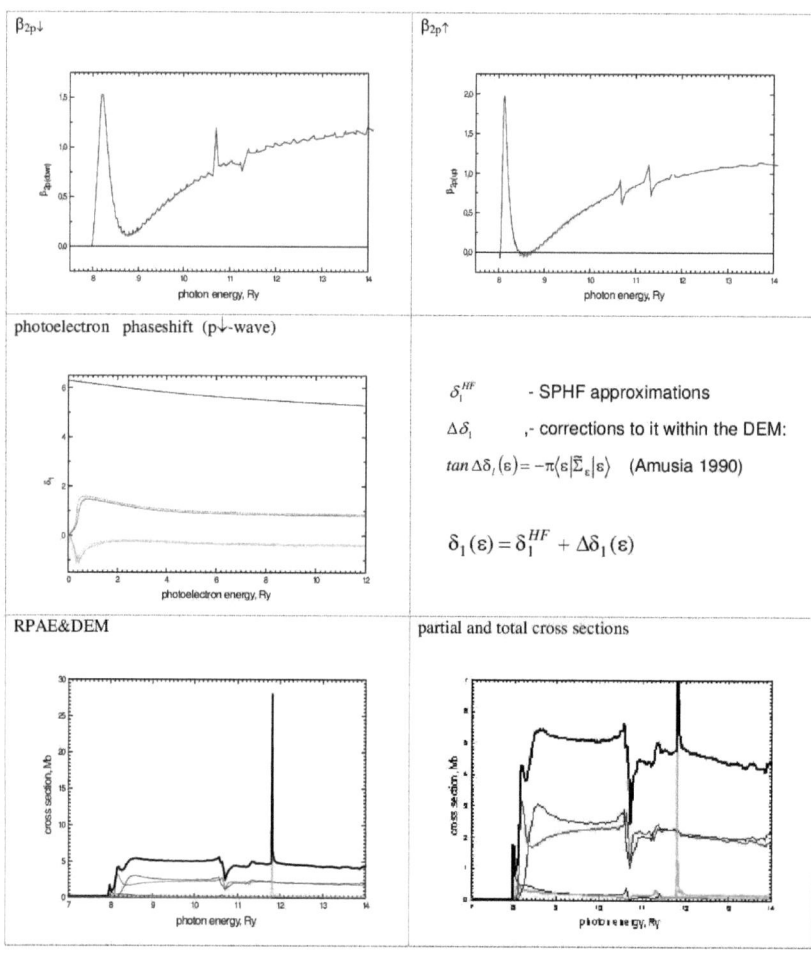

Figure 26. . The results of calculations within the DEM&RPAE method – the dynamical relaxation, polarization and interchannel interaction are included (see figure caption above).

0.020 (**r**- /∇ -forms). The dual nature of the "3p↓ shape-Feschbach" resonance is evident also from the complex behavior of a such sensitive to profiles form characteristics as the parameters of angular anisotropy $\beta_{2p\uparrow}$ and $\beta_{2p\uparrow}$ given in the upper panel of fig.26. The additional peculiarities of the Fano-profile type appear as a result in the total photodetachment cross section (low panel of fig.26). However, the total Si⁻ photodetachment crosssection in the energy region under investigation is dominated by the strong shape resonance peak at the 2s threshold. As in the case of C⁻ we have to conclude from the above analyses that the dynamical relaxation is the most pronounced effect in this strong correlated system.

Despite of the significant progress achieved in the merged ion beam-photon beam technique (Kjeldsen 2006, Müller 2015) demonstrated, in particularity, by the C⁻ (Walter *et al* 2006) and B⁻ (Berrah *et al* 2007) inner-shell photodetachment experiments, there is no experimental data on photoabsorption in the 2s inner shell region for Si⁻ up to date. Even keeping in mind that the data on the total photodetachment cross section could be extracted only from the analysis of high resolution spectra for the positive ion products of the related Auger-decay channels, the experimental evidence for the prominent resonance structure predicted in this work (fig.26) could be of a great interest.

The results of Chapter published for the first time.

12. Future trends

The further development of the presented many-body theory methods, of course, can be expected in the future. This Chapter is just to dwell upon some prospects of our program. One can notice the following feature of certain of the spectra presented above. Since an outer shell of negative ions is diffuse, a very rapid increase of the cross section for the channel np→εs (He⁻, B⁻, C⁻) with energy is typical of outer electron photodetachment. The corresponding partial cross sections give the main contribution to the total photoabsorption in the immediate vicinity of the outer shell treshould. The discrepancy between the cross section in r- and ∇-forms just above the first detachment limit happens to be appreciable. It might be diminished by taking into account the remainder of the correlational corrections to the electron-photon interaction vertex:

(23)

These corrections are most significant at low photon energies. In calculating of them a problem arises: singularities in the free-free dipole matrix element (Korol 1993) should be properly evaluated. The diagram (23) is important also from the theoretical point of view since its account within the DEM&RPAE method implies a full renormalization of a photon-electron interaction vertex, that makes the method self-consistent in in term of many-body theory.

The diagrams series, where a black circle is a photon interaction vertex presented the free-free dipole matrix elements obtained from a solution of the DEM&RPAE equation system like (13) and reducible self-energy part $\tilde{\Sigma}$ includes dynamical relaxation corrections,

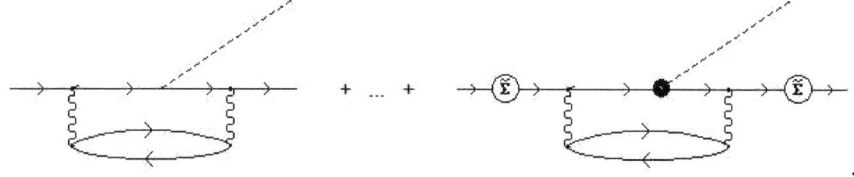

,

gives the idea of an application DEM&RPAE approach to the bremsstrahlung problem.

Summary

The new many-body theory method referred by us as RPAE&DEM is developed and applied to the study of photodetachment for a series of open-shell negative ions with emphasis on resonant features. The theme provides evidence enough to judge the potential of theoretical approaches for description of electronic correlations, because all the problems pertaining to negative ions tend to be essentially many-body problems. Collective phenomena happen to be of particular importance when photoprocesses occur at the energies which are close to the system natural frequency values - resonances.

The RPAE&DEM approach uses the Hartree-Fock approximation (HF) as a zero-order one. Many-electron correlations are incorporated within the Random Phase Approximation with Exchange, RPAE (to describe the dynamic collective response of an atomic system on to the external field), the Dyson Equation Method (DEM, to correct electron behaviour due to polarization and relaxation effects). Besides the Many-Body Perturbation Theory (MBPT) methods are applied depending on the role played by individual corrections. With the Feynman-Goldstone diagram technique one may identify certain classes of diagrams from the perturbation expansion and clarify the physical meaning of the included effects. Introducing physically meaningful corrections step by step we are able to analyze their influence on a photodetachment process.

By and large, negative ions structure and character of photodetachment processes are controlled by polarization of a neutral atom induced by an extra electron. Within the DEM the dynamic polarization potential including electron exchange is taken into account by an irreducible self-energy part of the single-particle Green function Σ which is calculated *ab initio* in the second order of the perturbation theory. The present modification of the DEM enables us to bind the outer electron in a negative ion and to correct the wavefunction and the single-particle energy of any electron, inner as well as outer, to take account for the polarization effects on a detaching electron, and thus to consider the influence of polarization interaction between a photoelectron and a core on the photodetachment process.

In studies of detachment from inner shells the atom-core relaxation is one of the decisive factor affecting the motion of a photoelectron. The presence of an electron vacancy causes the electronic cloud to deform, rearrange, with the outermost orbit "caving in". Static relaxation can be considered within our method without adding complexity but for a dynamic character of the process to be taken into account the new approach is developed within the DEM.

The DEM&RPAE is the many-body theory method for simultaneous inclusion of the dynamic polarization potential generated in the system "core + electron", the dynamic relaxation (screening) and the corrections within the scope of the RPAE.

This approach is applied, in particular, for the calculations of the total and partial outer-shells photodetachment cross sections, angular distribution parameters, photoelectron phaseshifts for the He^-, Cr^-, B^-, C^-. The subjects of investigation chosen are ions with open shells. Due to electronic cloud diffusion and possibilities for the formation of quasi-bound states in phototransitions from inner shells to vacant states the important role of collective effects is inherent in photodetachment from these. It was found that within the RPAE we can make some qualitative conclusions conserving photodetachment channels interference features, but for an adequate description of interference and resonance profiles it is necessary to take proper account of further collective phenomena. The parameters of shape- and window-resonances observed in photoabsorption spectra of the ions under discussion turn out to be very sensitive to introducing the corrections for relaxation and polarization. The concrete results and comparison with other theoretical predictions and experimental data are presented.

Inner-shell photodetachment of negative ions stand out as extremely sensitive probe and theoretical test-bed for important effects of electron-electron interaction because of the weak coupling between photon and target electrons. Many-body effects play a pronounced role here not only between the outer electron but also between the inner-shell electrons and outgoing electron. For our theoretical investigation of K-shell photodetachment we have chosen especially strong correlated system: the negative ions with an open p-shell C$^-$ and Si$^-$. The strong near-threshold resonance in the experimental high resolution spectra for the photodetachment from inner 1s-shell of the carbon negative is investigated. The complex "shape-Feshbach" nature of the resonance is revealed. The collective character of the response to the external electromagnetic field in the strongly correlated C$^-$ target is clearly demonstrated, with the dynamical relaxation of the core being the most pronounced of the collective effects. The analogical resonance structure is predicted for 2s,2p inner shell photodetachment of Si$^-$. The system is even more complex compared to C$^-$ since the interchannel interaction between the close inner shells should be considered. The whole atomic system turns out to be involved in resonant photodetachment process, core electrons and outgoing photoelectron being strongly coupled by correlation. Being a perfect test-bed for the many-body theory method applied, the challenge has demonstrated the approach ability of accurate inclusion of collective effects in its complex interplay.

To conclude, the collective effects are of considerable importance in the photodetachment from negative ions, especially with resonance processes. The concrete examples discussed above provide a convincing demonstration of this. The many-body theory methods are a reasonable powerful tool for the study of systems with inherent strong correlation interaction and show promise of precise calculations for photodetachment characteristics. These are very informative since the investigation of the dynamic of negative ions gives valuable insight into the fundamental problem of collective effects in atomic physics.

REFERENCES

Amusia M Ya and Cherepkov N A 1975 *Case Stud.At.Phys.* **5** 47.

Amusia M Ya and Chernysheva L V 1983 *Automated System for Atomic Structure Investigation* (Leningrad, in Russian).

Amusia M Ya, Dolmatov V K and Ivanov V K 1983 *JETPh* **1** 115-123.

Amusia M Ya 1990 *Atomic Photoeffect* (Plenum Press).

Andersen T 2004 *Phys.Rep.Rev.* **394** 157-313.

Balling P, Kristensen P, Stapelfeldt H, Andersen T and Haugen H K 1993 *J.Phys.B:At.Mol.Opt.Phys.* **26** 3531-39.

Berrah N, Bozek J D, Wills A A, Turri G, Zhou H L, Manson S T, Rude B, Gibson N D, Walter C W, VoKy L, Hibbert A 2001 *Phys.Rev.Lett.* **87** 253002.

Berrah N, Bozek J D, Turri G, Akerman G, Rude B, Zhou H L, and Manson S T 2002 *Phys.Rev.Lett.* **88** 093001.

Berrah N, Bozek J D, Bilodeau R C, *et al.* 2004a *Rad.Phys.Chem.* **70** 57-82.

Berrah N, Bilodeau R C, Ackerman G, *et al.* 2004b *Rad.Phys.Chem.* **70** 491-500.

Bilodeau R C, Bozek J D, Aguilar A, Ackerman G D, Turri G and Berrah N 2006 *Phys.Rev.Lett.* **93** 193001.

Bilodeau R C, Bozek J D, Aguilar A, Ackerman G D, Turri G and Berrah N 2006 *Phys.Rev.A* **73** 034701.

Bunge A V and Bunge C F 1979 *Phys.Rev.* **A 19** 452.

Chernysheva L V, Gribakin G F, Ivanov V K and Kuchiev M Yu 1988 *J.Phys.B:At.Mol.Opt. Phys.* **21** L419.

Covington A M, Aguilar A, Davis V T, Alvarez I, Bryant H C, Cisneros C, Halka M, Hanstorp D, Hinojosa G, Schlachter A S, Thompson J S, and Pegg D J 2001 *J.Phys.B: At.Mol.Opt.Phys.* **32** L735-L740.

Feigerle C S, Corderman R R, Bobashev S V and Lineberger W C 1981 *J.Chem.Phys.* **74** 1580.

Gribakin G F, Gribakina A A, Gul'tsev B and Ivanov V K 1992 *J.Phys.B:At.Mol.Opt.Phys.* **25** 1757-72.

Gribakin G F, Gul'tsev B V, Ivanov V K and Kuchiev M Yu 1990 *J.Phys.B:At.Mol.Opt.Phys.* **23** 4505.

Gorczyca T W 2004 *Rad.Phys.Chem.* **70** 407-415.

Haeffler G, Hanstorp D, Kiyan I Yu, Ljungblad U, Andersen H H and Andersen T 1996 *J.Phys.B:At. Mol.Opt.Phys.* **29** 3017-22.

Hazi A V and Reed K 1981 *Phys.Rev.* **A 24** 2269.

Hodges R V, Coggiola M J and Peterson J R 1981 *Phys.Rev.* **A 23** 59.

Ivanov V K, Kashenock G Yu, Gribakin G F, Gribakina A A 1996a *J.Phys.B:At.Mol.Opt.Phys.* **29** 2669.

Ivanov V K, Krukovskaya L P and Kashenock G Yu 1996b *J.Phys.B: At.Mol.Opt.Phys.* **29** L313.

Ivanov V K, Krukovskaya L P and Kashenock G Yu 1998 *J.Phys.B: At.Mol.Opt.Phys.* **31** 239.

Ivanov V K 1999 *J.Phys.B:At.Mol.Opt.Phys.* **32** R67-R101.

Ivanov V K and Kashenock G Y 1999 *P.Soc.Photo-Opt.Ins* **3687** 90-101.

Ivanov V K 2004 *Rad.Phys.Chem.* **70** 345-370.

Kashenock G Yu and Ivanov V K 1997 *J.Phys.B: At.Mol.Opt.Phys* **30** 4235.

Kashenock G Yu and Ivanov V K 1998 *Phys.Lett.A* **245** 110.

Kashenock G Yu and Ivanov V K 2006 *J.Phys.B: At.Mol.Opt.Phys* **30** 1379.

Kim D S, Zhou H L, and Manson S T 1997 *J.Phys.B: At.Mol.Opt.Phys.* **30** L1-L7.

Kjedsen H, Andersen P, Folkmann F, Kristensen B and Andersen T 2001 *J.Phys.B: At.Mol.Opt.Phys.* **34** 353-7.

Kjeldsen H 2006 *J.Phys.B: At.Mol.Opt.Phys* **39** R325.

Korol A V 1993 *J.Phys.B:At.Mol.Opt.Phys.* **26** 4769.

Kristensen P, Andersen H H, Balling P, Steele L D and Andersen T 1995 *Phys.Rev.A* **52** 2847.

Kutzner M 2004 *Rad.Phys.Chem.* **70** 95-104.

Le Dourneuf M, Niang C and Zeippen C J 1995 *5th EPS Conference on Atomic and Molecular Physiscs. Contributed papers* (Edinburgh, April 3-7, 1995) Ed R C Thompson, 800.

Lee D H, Tang C Y, Thompson J S, Brandon W D, Ljungblad U, Hanstorp D, Pegg D J, Dellwo J and Alton G D 1995 *Phys.Rev.A* **51** 4284.

Müller A 2015 *Phys.Scr.* **90** 054004

Nicolaides C A, Komminos Y and Beck D R 1981 *Phys.Rev.* **A 24** 1103.

Pegg D J 2004 *Rep.Prog.Phys.* **67** 857-905.

Peterson J R, Bae Y K and Huestis D L 1985 *Phys.Rev.Lett.* **55** 692

Radtsig A A and Smirnov B M 1986 *Parameters of Atoms and Atomic Ions*, Moscow (in Russian).

Ramsbottom C A, Bell K L and Berrington K A 1993 *J.Phys.B:At.Mol.Opt.Phys.* **26** 4399.

Ramsbottom C A and Bell K L 1995 *J.Phys.B:At.Mol.Opt.Phys* **28** 4501.

Ramsbottom C A and Bell K L 1999 *J.Phys.B: At.Mol.Opt.Phys.* **30** 1315-1333.

Saha H P and Compton R N 1990 *Phys.Rev.Lett.* **64** 1510 Tompson J S, Pegg D J, Compton R N and Alton G D 1990 *J.Phys.B:At.Mol.Opt.Phys* **23** L15.

Seman M and Branscomb L M 1962 *Phys.Rev.* **125** 1602.

Sobel'man I I 1963 *Introduction to Theory of Atomic Spectra* (Moscow, in Russian).

Walter C W, Seifert J A and Peterson J R 1994 *Phys.Rev.A* **50** 2257.

Walter C W, Gibson N D, Bilodeau R C, Berrah N, Bozek J D, Ackerman G D and Aguilar A 2006 *Phys.Rev.A* **73** 062702.

Zhou H L, Manson S T, Hibbert A, *et al.* 2004 *Phys.Rev.A* **70** 022713.

Xi JH and Fischer CF 1996 *Phys.Rev.A* **53** 3169-3177.

APPENDIX I (A)

FIVE YEARS OF STUDIES ON INNER-SHELL PHOTODETACHMENT

2001 THE FIRST INNER-SHELL PHOTODETACHMENT EXPERIMENTS
Kjeldsen H, Andersen P, Folkmann F, et al.
Inner-shell photodetachment of **Li-**
JOURNAL OF PHYSICS B 34 (10): L353-L357 MAY 28 2001

Berrah N, Bozek JD, Wills AA, et al.
K-shell photodetachment of **Li-**: Experiment and theory
PHYSICAL REVIEW LETTERS 87 (25): Art. No. 253002 DEC 17 2001

Covington AM, Aguilar A, Davis VT, et al.
Correlated processes in inner-shell photodetachment of the **Na**- ion
JOURNAL OF PHYSICS B 34 (22): L735-L740 NOV 28 2001

2002
Berrah N, Bozek JD, Turri G, et al.
K-shell photodetachment of **He-**: Experiment and theory
PHYSICAL REVIEW LETTERS 88 (9): Art. No. 093001 MAR 4 2002

Kjeldsen H, Andersen P, Folkmann F, et al.
Experimental study of 4f wavefunction contraction: 4d-photoionization of low-charged ions of **I**, Xe, Cs and Ba
JOURNAL OF PHYSICS B 35 (13): 2845-2860 JUL 14 2002

2003
Gibson ND, Walter CW, Zatsarinny O, et al.
K-shell photodetachment from **C**⁻: Experiment and theory
PHYSICAL REVIEW A 67 (3): Art. No. 030703 MAR 2003

2004
Gorczyca TW
Inner-shell photodetachment dynamics
RADIATION PHYSICS AND CHEMISTRY 70 (1-3): 407-415 MAY-JUN 2004

Berrah N, Bozek JD, Bilodeau RC, et al.
Studies of complex systems: from atoms to clusters
RADIATION PHYSICS AND CHEMISTRY 70 (1-3): 57-82 MAY-JUN 2004
(B^-, B_2^-, B_3^-)

Kjeldsen H, Folkmann F, Jacobsen TS, et al.
Feshbach resonances in inner-shell photodetachment: The case of **Te**⁻
PHYSICAL REVIEW A 69 (5): Art. No. 050501 MAY 2004

Bilodeau RC, Bozek JD, Aguilar A, et al.
Photoexcitation of He⁻ hollow-ion resonances:observation of the $2s2p^2$ ⁴P state
PHYSICAL REVIEW LETTERS 93 (19): Art. No. 193001 NOV 5 2004

2005
Berrah N, Bilodeau RC, Bozek JD, et al.
Double photodetachment in He-: Feshbach and triply excited resonances
JOURNAL OF ELECTRON SPECTROSCOPY AND RELATED PHENOMENA 144: 19 JUN 2005

APPENDIX I (B)

FIVE YEARS OF STUDIES ON INNER-SHELL PHOTODETACHMENT

Li⁻

Zhou HL, Manson ST, VoKy L, et al.
Dramatic structure in the photodetachment of inner shells of negative ions: Li⁻
PHYSICAL REVIEW LETTERS 87 (2): Art. No. 023001 JUL 9 2001
the enhanced version of the **R-matrix** code, with an upgraded asymptotic part to handle the negative ion system

EXPERIMENT (Kjeldsen et al. 2001),
EXPERIMENT / R-matrix CALCULATIONS (Berrah et al. 2001)

Gorczyca TW, Zatsarinny O, Zhou HL, et al.
Postcollision recapture in the K-shell photodetachment of Li⁻
PHYSICAL REVIEW A 68 (5): Art. No. 050703 NOV 2003
R-matrix ;
a fully quantum mechanical approach for treating **the post-collisional interaction effect, with a Multichannel Quantum Defect R-matrix method modified with an optical potential**

Ivanov VK et al.
Electron photodetachment from the 1s shell of a negative lithium ion
TECHNICAL PHYSICS LETTERS 29 (8): 620-623 2003
Many-Body Perturbation Theory method

He⁻

Xi JH, Fischer CF
Photodetachment cross section of He- (1s2s2p P-4(0)) in the region of the 1s detachment threshold
PHYSICAL REVIEW A 59 (1): 307-314 JAN 1999
Multi-Configuration Hartree-Fock (MCHF)

Zhou HL, Manson ST, Ky LV, Hibbert A, Feautrier N
Photodetachment of **He**- 1s2s2p P-4(0) in the region of the 1s threshold
PHYSICAL REVIEW A 64 (1): Art. No. 012714 JUL 2001
R-matrix

EXPERIMENT / R-matrix CALCULATIONS (Berrah et al. 2002)

Zatsarinny O, Gorczyca TW, Fischer CF
Photodetachment of He- 1s2s2p (4)p(o) in the region of the 1s threshold
JOURNAL OF PHYSICS B 35 (20): 4161-4178 OCT 28 2002
the theory employs an R-matrix calculations with a spline basis

Sanz-Vicario JL, Lindroth E
Resonant triply excited states in the photodetachment of He- 1s2s2p P-4(o)
PHYSICAL REVIEW A 65 (6): Art. No. 060703 JUN 2002
complex scaled configuration interaction CS CI

Sanz-Vicario JL, Lindroth E, Brandefelt N
Photodetachment of negative helium ions below and above the 1s ionization threshold: A complex scaled configuration-interaction approach
PHYSICAL REVIEW A 66 (5): Art. No. 052713 NOV 2002
CS CI with the PCI recapture included semiclassically

APPENDIX I (C)

REVIEWS

Andersen T
Atomic negative ions: structure, dynamics and collisions
PHYSICS REPORTS-REVIEW SECTION OF PHYSICS LETTERS 394 (4-5): 157-313 MAY 2004

Pegg DJ
Structure and dynamics of negative ions
REPORTS ON PROGRESS IN PHYSICS 67 (6): 857-905 JUN 2004

Ivanov VK
Many-body effects in negative ion photodetachment
JOURNAL OF PHYSICS B-ATOMIC MOLECULAR AND OPTICAL PHYSICS 32 (12): R67-R101 JUN 28 1999

Radiation Physics and Chemistry
Volume 70, Issues 1-3, Pages 1-463 (May - June 2004)
Photoeffect: Theory and Experiment
Edited by R.H. Pratt and S.T. Manson

Berrah N, Bozek JD, Bilodeau RC, et al.
Studies of complex systems: from atoms to clusters
RADIATION PHYSICS AND CHEMISTRY 70 (1-3): 57-82 MAY-JUN 2004

Berrah N, Bilodeau R C, Ackerman G, et al.
RADIATION PHYSICS AND CHEMISTRY 70 (1-3): 491 MAY-JUN 2004

Gorczyca TW
Inner-shell photodetachment dynamics
RADIATION PHYSICS AND CHEMISTRY 70 (1-3): 407-415 MAY-JUN 2004\

Lindroth E, Sanz-Vicario JL
Photodetachment of few-electron negative ions
RADIATION PHYSICS AND CHEMISTRY 70 (1-3): 387-405 MAY-JUN 2004

Ivanov VK
Theoretical studies of photodetachment
RADIATION PHYSICS AND CHEMISTRY 70 (1-3): 345-370 MAY-JUN 2004

EXPERIMENT

Kjeldsen H, Folkmann F, van Elp J, et al.
Absolute measurements of photoionization cross-sections for ions
NUCLEAR INSTRUMENTS & METHODS IN PHYSICS RESEARCH SECTION B-BEAM INTERACTIONS WITH MATERIALS AND ATOMS 234 (3): 349-361 JUN 2005

Negative ions : stored in ASTRID

H^- D^- $^3He^-$ $^4He^-$ $^9Be^-$ $^{11}B^-$ $^{12}C^-$ $^{16}O^-$ $^{19}F^-$ $^{28}Si^-$

$^{32}S^-$ $^4He_2^-$ $^{12}C_{2-8}^-$ OH^- Al_{1-12}^- C_{48-60}^- C_{70}^-

http://www.isa.au.dk/research/research.html#SR
http://www.ifa.au.dk/amo/atomphys/atomphys.htm

APPENDIX II

MAIN RELATED PUBLICATIONS

1. V.K.Ivanov, G.Yu.Kashenock, G.F.Gribakin and A.A.Gribakina "2s,2p photodetachment from the $He^-(^4P^0)$ negative ion within the Dyson Equation method". *J.Phys.B: At.Mol.Opt.Phys.* **29** 2669-2687 (1996). doi:10.1088/0953-4075/29/13/007

Abstract. The results of many-body calculations of the 2s, 2p $He^-(^4P^0)$ photodetachment cross section are presented. Many-electron correlations, in particular the polarization interaction between a photoelectron and atomic core electrons in both initial and final states, are taken into account within the Dyson equation method. The calculated binding energies for 2s and 2p electrons are found to be equal to 1.075 and 0.080 eV, respectively. The 2p photodetachment cross section agrees quite well with available experimental data and previous calculations. The 2s photodetachment cross section reveals the well known strong shape resonance just above the 2^3P threshold, which is associated with a $1s2p^2$ quasi-bound state, although the calculated peak value is less than the experimental one.

2. V.K.Ivanov, L.P.Krukovskaya and G.Yu.Kashenock "Near-threshold shape resonance in Cr^- outer-shell photodetachment". *J.Phys B: At.Mol.Opt.Phys.* **29** L313-L319 (1996). doi:10.1088/0953-4075/29/8/002

Abstract. The results of many-body calculations for outer 4s-subshell photodetachment of the negative ion Cr^- $(...3d^54s^2$ $^6S)$ are presented. The 4s cross section reveals a near-threshold strong maximum, which may be attributed to the presence of a quasi-bound `4p' state in the continuum. It is shown that the resonance depends strongly on core-polarization effects. The dynamical polarization interaction between the atomic core and the extra electron have been taken into account within the Dyson equation method. The latter was used to bind the extra electron to a neutral Cr atom in the 4s state, to obtain its energy and wavefunction (the calculated energy is 0.782 eV, the electron affinity is 0.667 eV) and to correct the photoelectron wavefunctions. The wavefunctions thus obtained are used for the photodetachment cross section calculations.

3. G.Yu.Kashenock and V.K.Ivanov "Collective effects in B^- photodetachment" *J.Phys.B: At.Mol.Opt.Phys.* **30** 4235-4253 (1997). doi:10.1088/0953-4075/30/19/014

Abstract. A study of the structure and photodetachment of the $B^-(...2s^22p^2$ $^3P)$ negative ion has been performed using many-body theory. An electron affinity for the $B(^2P^o)$ of 0.273 eV was calculated within the Dyson equation method. A new approach for concurrent consideration of intrachannel and interchannel interactions and dynamic-core polarization and relaxation (screening) effects has been developed to calculate the total and partial photodetachment cross sections. The B^- photoabsorption spectrum reveals a complex interference structure due to correlations between the two outer subshells, primarily strong interaction with the $...2s2p^3$ ^3D shape resonance. The influence of many-electron dynamic corrections on the photodetachment process is discussed. The resonance and interference profiles have been shown to be very sensitive to the proper account of collective phenomena. Reasonable agreement between the calculations and experimental data has been achieved.

4. V.K.Ivanov, L.P.Krukovskaya and G.Yu .Kashenock "The evidence of giant "$3p^5 3d^6 4s^2$" resonances in Cr⁻ photodetachment" *J.Phys.B:At.Mol.Opt.Phys.* **31** 239-247 (1998). doi:10.1088/0953-4075/31/2/007

Abstract. The first ab initio calculations of the refined Cr⁻ *ion energy structure through consideration of the interelectron interactions and the photodetachment cross section (partial and total) in the vicinity of the $3p^6$ shell threshold are reported. The resonance behaviour is found to be associated with the excitation and autodetachment of the quasibound "$3p^5 3d^6 4s^2$" state to 3p, 3d and 4s channels.*

5. G.Yu.Kashenock and V.K.Ivanov. "The $2s^1 2p^4$ autodetachment resonance in the C⁻ negative ion". *Phys.Lett.A* **245**, 110-116 (1998). doi:10.1016/S0375-9601(98)00330-2

Abstract. The calculations of the photodetachment cross section for the C⁻ negative ion has been performed within the newly-developed many-body theory method, the RPAE interchannel interaction and dynamic relaxation and polarization corrections being included. The $2s^1 2p^4$ shape resonance is shifted to a higher photon energy and broadens as compared to the resonance parameters determined earlier within the RPAE, which is consistent with the experimental evidence and the recent R-matrix calculations.

6. G.Yu.Kashenock and V.K.Ivanov. "Inner-shell resonant photodetachment of C⁻: the dramatic role of dynamical relaxation in collective response". *J.Phys.B: At.Mol.Opt.Phys.* **39** 1379-93 (2006). doi:10.1088/0953-4075/39/6/010

*Abstract. The near-threshold strong resonance in 1s inner-shell photodetachment of the carbon negative ion is investigated and the role of collective effects in its formation is evaluated. The shape resonance parameters, the resonance energy of $\varepsilon_{res} = 0.857$ eV and width of $\Gamma = 0.122$ eV, are determined using the many-body theory (RPAE&DEM) method. The complex mixed ('shape-Feshbach') nature of the resonance is revealed. The calculated cross sections are in fair agreement with the recent experiment (Gorczyca T W 2004 Rad. Phys. Chem. **70** 407–15). The collective character of the response to the external electromagnetic field in the strongly correlated C⁻ target is clearly demonstrated. The dynamical relaxation of the core is the most pronounced feature in the interplay of collective effects.*

Printed by Books on Demand GmbH, Norderstedt / Germany